General editor: Graham Hand

Brodie's Notes on Gera...

My Family and Other Animals

Kenneth Hardacre MA

Pan Books London, Sydney and Auckland

First published by James Brodie Ltd

First published 1979 by Pan Books Ltd

This revised edition published 1991 by
Pan Books Ltd, Cavaye Place, London SW10 9PG

9 8 7 6 5 4 3 2 1

© Kenneth Hardacre 1991

ISBN 0 330 50315 4

Photoset by Parker Typesetting Service, Leicester

Printed and bound in Great Britain by
Clays Ltd, St Ives plc, Bungay, Suffolk

This book is sold subject to the condition that it shall not,
by way of trade or otherwise, be lent, re-sold, hired out
or otherwise circulated without the publisher's prior consent
in any form of binding or cover other than that in which
it is published and without a similar condition including this
condition being imposed on the subsequent purchaser

Contents

Preface by the general editor v

The author and his work 1

Chapter summaries, textual notes, and revision questions 12

Characters and creatures
Gerry 37, Mother 38, Larry 39, Leslie 40, Margo 41, Spiro 41, Theodore 42, Kralefsky 42, Achilles 43, Ulysses 44, Geronimo 44, Cicely 44, The Magenpies 45, Old Plop 45, Alecko 45

Setting and structure 46

Style 49

General questions and sample answer in note form 59

Further reading and reference 63

Page references in these Notes are to the standard Penguin edition of *My Family and Other Animals*, but as references are also given to Parts and Chapters, the Notes may be used with any edition of the book.

The quotations from *Prospero's Cell* appear with acknowledgements to the author and the publishers Messrs Faber and Faber.

Preface by the general editor

The intention throughout this study aid is to stimulate and guide, to encourage your involvement in the book, and to develop informed responses and a sure understanding of the main details.

Brodie's Notes provide a clear outline of the play or novel's plot, followed by act, scene, or chapter summaries and/or commentaries. These are designed to emphasize the most important literary and factual details. Poems, stories or non-fiction texts combine brief summary with critical commentary on individual aspects or common features of the genre being examined. Textual notes define what is difficult or obscure and emphasize literary qualities. Revision questions are set at appropriate points to test your ability to appreciate the prescribed book and to write accurately and relevantly about it.

In addition, each of these Notes includes a critical appreciation of the author's art. This covers such major elements as characterization, style, structure, setting and themes. Poems are examined technically – rhyme, rhythm, for instance. In fact, any important aspect of the prescribed work will be evaluated. The aim is to send you back to the text you are studying.

Each study aid concludes with a series of general questions which require a detailed knowledge of the book: some of these questions may invite comparison with other books, some will be suitable for coursework exercises, and some could be adapted to work you are doing on another book or books. Each study aid has been adapted to meet the needs of the current examination requirements. They provide a basic, individual and imaginative response to the work being studied, and it is hoped that they will stimulate you to acquire disciplined reading habits and critical fluency.

Graham Handley 1991

The author and his work

Gerald Malcolm Durrell is a zoologist, an animal collector, a writer and lecturer, a regular contributor to BBC sound and television programmes, the founder and honorary Director of the Jersey Zoological Park, later the Jersey Wildlife Preservation Trust, and Founder-Chairman of the Wildlife Preservation Trust International.

He was born in 1925 in Jamshedpur, an industrial city near Calcutta in north-east India, where his father was a civil engineer. He was the youngest of four children, and when his father died in 1928 his mother brought her family back to England and settled for a time in Bournemouth. From the age of two Gerald had been keen on animals. 'According to my parents,' he says, 'the first word I was able to say with any clarity was not the conventional "Mamma" or "Dadda", but the word "Zoo", which I would repeat over and over again in a shrill voice until someone, in order to shut me up, would take me to the zoo.' When he went to a kindergarten, an establishment had to be found where young Gerald would be allowed to take some of his pets to school with him. In 1933 the family went to live on the Continent, and he was educated by private tutors in France, Italy and Switzerland.

In 1934, after a brief return to Bournemouth, they eventually settled on the island of Corfu, off the coast of Albania, where they lived until 1939. *My Family and Other Animals* is an account of these five years. During this time Gerald made a special study of zoology and, though he had several private tutors, he says he owes most to a friend of his eldest brother Lawrence, Dr Theodore Stephanides, RAMC, who gave him immense help in his naturalist studies. 'I had a great number of pets, ranging from owls to sea horses, and I spent all my spare time exploring the countryside in search of fresh specimens to add to my collection.' Then came a return to Bournemouth and more tutors.

During World War II Gerald Durrell was engaged on agricultural research and the study of animal ecology. In 1945 he went for a year to Whipsnade Zoo as a student keeper. *Beasts in My Belfry* (1973) is a lively account of his time at Whipsnade and of the colleagues he worked with there, gaining experience of the larger animals 'which were not so easy to keep at home' – lions, tigers, bears, zebras, buffalo, wolves, yaks, gnus and the very rare Père David deer. It was while he was working at Whipsnade, he says, that he realized the full meaning of the term 'rare' and conceived the idea that, should he ever acquire a zoo of his own, its main function would be the breeding and conservation of rare species.

When he was twenty-one Durrell inherited some money from a great-aunt and this, together with what he had earned as a student keeper, enabled him to finance his first trip to collect animals and bring them back alive for various zoos; and ever since he has been going out regularly to distant places where such animals live, writing an account of each trip in order to raise money for financing the next one.

His first expedition was a six-month collecting trip with John Yealland to the British Cameroons in West Africa in 1947. Of this he wrote an account in his first book, *The Overloaded Ark* (1953). He describes with remarkable sympathy the native hunters with whom he worked, catching porcupines, crocodiles, chameleons and monkeys; and he has the ability to arouse our interest in the individual snakes, bats, rats and toads he collected. We share the thrill of coming upon the rare angwantibo; the delights of George, the baboon who attended a local dance; the difficulties encountered in climbing N'da Ali, the mountain with a *ju-ju*, or magic spell; and the fascinating story of the life and death of Cholmondeley the chimpanzee.

In 1948 came a second expedition to the Cameroons, this time with Kenneth Smith. They made their headquarters at Bafut, where they became great friends of the Fon, the charming local native ruler, who lent them a villa and supplied them with four native hunters and a pack of six mongrel dogs. These Durrell called 'the Bafut beagles' and with them he hunted rats, booming squirrels, the dwarf mongoose and snakes of all kinds – including the much dreaded but really harmless Que-fong-goo and what he supposed to be a typhops (a harmless blind-snake) which turned out to be a poisonous one whose bite nearly proved fatal. During a night of festivities Durrell taught the Fon, his advisers and his many wives, how to dance the conga; then at dawn he set off into the forest to catch his first hairy frog. Durrell tells the story of the whole expedition in *The Bafut Beagles* (1954), which also includes an account of his activities at another camp at Mamfe and describes a hunt for flying mice.

A few months later Durrell and Kenneth Smith set off on another collecting trip, this time to British Guiana. This is described in *Three Singles to Adventure* (1954), which tells of their encounters with lizards, snakes, squirrel monkeys, soldier rats, opossums, sloths and a curassow called Cuthbert. In the Rupununi savannah they caught a large cayman (a South American alligator), and had an amusing but exhausting chase after an anteater; and in the creeklands near Charity, Durrell was fortunate enough to find a specimen of the pipa toad, which hatches her young in thousands of little pockets on her back.

From 1950 to 1953 Durrell was busy writing and broadcasting. He married in 1951 Jacqueline Sonia Rasen, and in 1953 he and his wife went on a trip to Argentina and Paraguay. In the pampas near Buenos Aires they caught oven birds, burrowing owls, a comical baby screamer called Egbert, and two armadillos known as the Terrible Twins. Then they moved to the Chaco territory of Paraguay, where they found a species of armadillo that rolls itself into a ball resembling an orange; a douracouli monkey called Cai; a crab-eating raccoon which they christened Pooh; the rare Budgett's frog; Lindo, a beautiful fawn; a fer-de-lance, the most poisonous and bad-tempered of South American snakes; Sarah Huggersack, a baby giant anteater; and an anaconda nine feet long. Unfortunately a political revolution made it impossible for the Durrells to obtain transport, and most of their animals had to be released from their cages and driven, reluctantly, away. Back in Argentina they witnessed and filmed a rhea-hunt. 'We returned from South America,' says Durrell, 'with only a handful of specimens in place of the large collection we had hoped for. However, even a failure has its lighter side.' This Durrell managed to portray in his account of the trip, *The Drunken Forest* (1956).

During 1955 Durrell was filming in Cyprus, and in the following year he did his first television series and published *My Family and Other Animals* (1956).

Ever since he was quite small Durrell wanted to have his own zoo. After travelling all over the world catching wild animals for other people's zoos, he says, he found it heart-breaking to have to part with creatures he had spent six months looking after and had brought back successfully from strange parts of the world. Eventually he managed to save enough money for his own zoo and in 1957, though he had no idea where the zoo was to be, he set off with his wife on his third Cameroon expedition to collect his own animals. After an unsuccessful attempt to catch a python in a cave they went to Eshobi. Here they encountered the rare bird *Picathartes* and caught a needle-clawed lemur (or bushbaby). At Bafut, Durrell renewed his friendship with the Fon and introduced his wife to him. There was more dancing and the natives brought Durrell a fine collection of 'beef'. The whole story is told in A Zoo in my Luggage (1960) – amusing episodes with monkeys, with a baboon called Georgina, 'a creature of tremendous personality and with a wicked sense of humour', with a chimpanzee called Minnie, and with Cholmondeley St John, an aristocratic and highly intelligent ape. Here too are the sad stories of the clawed toads and of Bertram, the dormouse who lost his tail.

When Durrell and his wife returned to England the animals were first kept in the house and garden of his mother's home in Bournemouth, much to the consternation of the neighbours, and later in the basement of a large department store, where Georgina escaped and was only recaptured with much difficulty and after considerable damage to chandeliers and china. Attempts to find a site for the zoo were for a long time unsuccessful, until at last, in Jersey, they found a lovely old manor house, whose grounds would provide a perfect setting for the animals. After much construction work had been done, building cages and providing special heating arrangements for the more delicate animals, the Jersey Zoological Park at Les Augres Manor was opened in 1958. Here, Durrell says, he 'tries to keep the animals as tame as possible, so that they are always ready to show themselves to the visitors, and attempts to gather some of the rarer animals that are in danger of becoming extinct'. He has written an account of some of his animals, illustrated with many fine photographs by W. Suschitzky, in *Island Zoo* (1961), and in *Menagerie Manor* (1964) he describes a tour of the zoo, with an account of a typical day from dawn to darkness and of escapades, difficulties and successes with some of the more colourful creatures.

In 1958–9 Durrell and his wife spent eight months in Argentina in order to bring back a South American collection for their zoo. After protracted negotiations with customs officials they were at last able to start collecting. The story of their travels is told in *The Whispering Land* (1961). They made a trip to a penguin colony on the Patagonian coast, and then northwards to a colony of fur seals on the Valdes peninsula, where they were delighted to find the Darwin's rhea (the South American counterpart of the African ostrich) and the hairy armadillo, and where they came upon a colony of elephant seals almost by accident. At Jujuy they collected a wild cat, and many parrots and other birds. A hair-raising journey across flooded rivers brought them to Oran, where they bought more parrots, Brazilian rabbits, a pathetic ocelot, a puma from a circus, and a young peccary with endearing habits. On a trip into the mountains Durrell kept an unsuccessful all-night vigil for vampire bats but won a pigmy owl as a consolation prize. There are also visits to a number of remarkable human specimens – an exiled Hungarian cabinet minister, and a self-taught ornithologist who worked in a sawmill and whom everyone except Durrell regarded as mad. The book ends with an account of the trials and tribulations associated with bringing back a large collection of animals on a long sea voyage that began with the nursing of a peccary who had caught pneumonia.

In 1962 Durrell, Jacquie, a television producer and a cameraman were engaged in a six-month, 45,000-mile journey through New Zealand, Australia and Malaya. Their purpose was to see what was being done about the conservation of wildlife in these countries and to make a series of television films for the BBC. They met a fascinating variety of people and animals – among the latter many rare creatures, like the duck-billed platypus, the tuatara ('a genuine, living, breathing prehistoric monster') and oddly named ones, like the long-nosed bandicoot and the racquet-tailed drongoe – and had the exciting good fortune to come upon a rare opportunity to film the birth of a kangaroo. *Two in the Bush* (1966) is Durrell's account of the expedition.

On his return from this prolonged trip he found the Zoological Park almost bankrupt. After taking over the management of the place himself he set up the Jersey Wildlife Preservation Trust in 1964. This meant an extension of the aims of the original Zoological Park to include collecting, studying and breeding threatened species – in particular, rare primates – in order to ensure their ultimate survival in the world. Durrell believes that conventional methods of protection are not enough and that zoos should set up controlled breeding programmes. A key book for understanding his aims and ideals is *The Stationary Ark* (1976), which deals with all aspects of the proper purpose of a zoo, all illustrated with anecdotes and accounts presented in his own inimitable style.

Further collecting trips to Sierra Leone, in West Africa, for leopard cubs and Colobus monkeys, and to Mexico for volcano rabbits and thick-billed parrots are the subject of *Catch Me a Colobus* (1972).

Durrell has produced several books specially for younger readers. *The New Noah* (1955), written in response to many demands that he should tell some of his stories for them personally, deals with collecting in the Cameroons, hunts and captures in Guiana, and perambulations in Paraguay. *Encounters with Animals* (1958) is a collection of broadcast talks on different animal subjects – animal courtships, architects, inventors and warfare – and includes some descriptions of particular animals; two kusimanses known as the Bandits, Wilhelmina the whip-scorpion, Sarah Huggersack, the baby giant anteater, and Pavlo, the black-eared marmoset. There are, as well, two human animals, MacTootle, an Irish banana-planter in the Cameroons, and Sebastian, the lively ninety-five-year-old South American gaucho. *Look at Zoos* (1961) is an 'interest book' for young children. After mentioning zoos in the past (Chinese and Aztec), Durrell deals with some of the problems of running a zoo (providing the right kinds of cages, feeding the animals and accustoming them to a substitute food when their original diet is not available). He points out the sort of thing to look out for in a zoo – the different kinds of protective coloration in animals and their use of display in defence and courtship. He discusses the rearing of baby animals and fish, describes how animals discipline their young and explains the different ways in which various animals feed.

Durrell's talent for creative and imaginative writing is as impressive as his professional knowledge of animal behaviour and both are combined to advantage in the fiction he has written for younger readers.

The Donkey Rustlers (1968) is set on a Greek island where the inhabitants depend for their transport on donkeys. Here an English family, the Finchberry-Whites, rent for part of each year a large Venetian villa at Kalanero. Their children, David and Amanda, spend much of their time with their Greek friend,

Yani, who is in danger of losing his inheritance, confiscated by the malevolent mayor. By means of an ingenious plot, which involves kidnapping and hiding all the donkeys and blackmailing the mayor into offering a reward more than large enough to reimburse Yani, all ends well.

The Talking Parcel (1974) is the story of how Simon, Peter and Penelope, while on holiday in Greece, rescue from the sea a very large parcel containing Parrot and Dulcibelle, his housekeeper, who have been exiled from Mythologia. This is a wonderful underground country where an inventor called Hengist Hannibal Junketberry has preserved many species of mythological creatures. The country is in danger of being taken over by cockatrices but, with the help of the children and the concerted efforts of unicorns, griffins, werewolves, firedrakes and Ethelred, a cockney toad, Cockatrice Castle is successfully stormed and order is restored.

Durrell's *The Fantastic Flying Journey* (1987) is beautifully illustrated by Graham Percy. Great-uncle Lancelot has designed a navigable balloon with a bamboo house of several floors instead of the usual basket. In this he and the three Dollybutt children, Emma, Conrad and Ivan, set out round the world in search of Lancelot's lost brother, Perceval. A marvellous invention enables them to have conversations and interviews with all the animals they meet. In Central Africa they meet monkeys and Lancelot discusses conservation with a gorilla, and on the River Zambesi they watch crocodile eggs hatching. In Australia they visit Ayers Rock and meet a marsupial mouse, koala bears, a duck-billed platypus and, of course, kangaroos. From the North Pole, where they talk with polar bears, musk oxen and wolves, they travel through Canada and North America to the River Amazon and meet a boa constrictor, jaguars, tapirs and howler monkeys. Emma keeps a diary with sketches and paintings and notes on everything they meet and says, 'We've learned lot on this journey. We must make sure we use that knowledge to help wildlife when we get home.' After dangers and discoveries they do get home safe and sound – and find Perceval.

Rosy is My Relative (1968) is Durrell's first book of adult fiction. In a last wish from his uncle, young Adrian Rookwhistle is asked to look after Rosy and 'sustain her in the style to which she is accustomed'. Rosy arrives almost immediately and turns out to be an ex-circus elephant with a fondness for alcohol. Adrian decides to take her to the holiday coast, in the hope of finding a place for her in a circus. In the course of their journey together Rosy causes havoc to the Monkspepper Hunt, wrecks a birthday party and much of the home of Lord Fenneltree and brings chaos to a production of *Ali Baba and the Forty Thieves* and the theatre in which it is being performed. Adrian and Rosy eventually come face to face with the law. Durrell's account of the subsequent trial (and its surprising outcome) is a masterpiece of comic writing.

A second novel followed in 1981. In *The Mockery Bird* Peter Foxglove is the new Assistant to Hannibal Oliphant, Political Adviser to the King of Zenkali, an island paradise in the Pacific Ocean, just prior to its becoming self-governing. Its European inhabitants are an eccentric collection and of the two native tribes one worships the Fish God, the other the Mockery Bird. The latter no longer survives, the last specimens having been eaten up by French settlers, and of the species of Ombu tree only a unique, carefully tended example appears to have survived. High feelings are aroused by the threatened construction of an airstrip and the flooding of a valley for a hydro-electric plant. When Peter and his new friend Audrey, in their exploring, come upon a valley which has fifteen pairs of Mockery Birds and four hundred Umbu trees, the result is chaos and confusion, and a clash between conservation and commercial development which leads to a climax that is at once exciting, impressive and symbolic.

Two volumes of short stories contain material that can perhaps best be described as heightened recollection. The five pieces in *Fillets of Plaice* (1971) are first person accounts by Gerald: an eventful and trouble filled birthday sea trip by the Durrell family and friends from Corfu to the mainland; teenage Gerry's job as an assistant in a London pet shop and the various eccentrics this brings him into contact with; a three-week 'rest' from over-work and over-worry in a nursing home with a galaxy of nurses; and a meeting later in life with one of his most striking Bournemouth girlfriends. The British Cameroons – see the reference to *The Bafut Beagles* (p.2) – is the setting for a story of how, with Gerald's help, a young District Officer deals with an important visit from his District Commissioner, and of what goes wrong.

The Picnic and Suchlike Pandemonium (1979) comprises six short stories. Their subjects are: the Durrell family's disastrous picnic at Lulworth Cove, in Dorset; their incident-filled voyage on a peculiar Greek ship from Venice to Corfu, to revisit 'the scenes of our youth'; the farcical results of a case of adultery and its climax in Venice; how Durrell's reading *The Psychology of Sex* in a Bournemouth hotel leads to his being consulted by every member of the staff in turn on their sexual problems; a strange culinary tale from the proprietor of a Provençale hostelry; and an account of macabre events in a French château.

In 1972 Durrell founded the Wildlife Preservation Trust International and has been its Chairman ever since. Part of each year he devotes to writing books in order to fund his conservation work, travelling round the world to mount expeditions, make television documentaries and promote international co-operation on wildlife conservation issues.

Between 1976 and 1981 there were expeditions to Mauritius, Assam, Mexico and Madagascar. *Golden Bats and Pink Pigeons* (1977) is an account of two trips to Mauritius to discuss ways of helping the authorities there to save the many rare and endangered species still to be found there, and to catch and take back to Jersey specimens for breeding there as a safeguard – an undertaking that was successfully completed.

In 1979, after his first marriage was dissolved, Durrell married Lee Wilson McGeorge, a zoologist from Tennessee who has made a special study of vocal communication in birds and mammals, and of social behaviour in mammals. In 1983 Durrell was awarded an OBE for his conservation work and in 1988 he was awarded an Honorary Doctorate by the University of Durham.

Gerald and Lee Durrell collaborated in writing *The Amateur Naturalist* (1982). Subtitled 'a practical guide to the natural world', this classic work of reference on observing, understanding, recording, studying and helping to conserve nature, is full of information on the vegetation and living creatures of a variety of habitats, from meadows and hedgerows, wetlands and seashores, to deserts and tundras, mountains and tropical forests. Moreover, from time to time there are relevant autobiographical asides and reminiscences of his early days on Corfu, incidents in his travels in other parts of the world, and an exciting discovery during Lee's birthday party at their house in southern France. There is more about Leslie and about a particular hunting expedition on which young Gerry was allowed to accompany him, and an account of 'a very silly and dangerous enterprise' in which he photographed the nest of a griffon vulture twelve metres below the top of a cliff face.

A television series of thirteen programmes was based on the book and first shown in 1983. *How to Shoot an Amateur Naturalist* (1984) tells the story of the problems, successes and widely different settings (requiring some 49,000 miles of travelling) involved in making the series. The human beings Durrell met were as colourful as the vegetation and animal specimens and are presented in the same impressive, attractive and amusing prose. In a chapter on Corfu he revisits, after more than forty years, the 'Snow-White Villa' and recalls many of the striking things that happened there.

Gerald and Lee Durrell collaborated again in writing *Durrell in Russia* (1986), a large format, lavishly illustrated account of a ten-month visit to make a series of thirteen television programmes, filmed in nineteen different locations at different times of the year. Few countries in the world give as much importance to conservation as Russia and the Durrells met the staffs of several impressive nature reserves, but also many ordinary Russians – like the children of a school heavily orientated towards ecology and the shoppers of Samarkand in a market that was old in the time of Genghis Khan.

Durrell has made ten films for television and in addition the BBC made a ten-part television series, *My Family and Other Animals*, adapted from Durrell's book by Charles Wood and produced by Joe Waters. It was first broadcast in 1987 and has Brian Blessed as Spiro, Hannah Gordon as Mrs Durrell and Darren Redmayne as Gerry.

Finally, of Durrell's thirty or so books, mention must be made of two which have special interest for readers of *My Family and Other Animals*, since they form sequels to that work. In *Birds, Beasts and Relatives* (1969) there are new adventures on Corfu and we meet again, besides the family, many old friends – Spiro, Theodore, George, Lugaretzia and Kralefsky (who tells a story of his taking over from the lion tamer at a circus performance) – and many new ones, like Papa Demetrius of the olive press and Countess Mavrodaki, who gives Gerry a gargantuan lunch and a barn owl, which he christens Lampadusa ('He was, I thought, one of the most beautiful birds I had ever seen'). There are, of course, familiar animals too, like Ulysses, the Scops owl, and

Roger, the dog ('that indefatigable student of natural history'), who goes everywhere with Gerry.

The Garden of the Gods (1978) contains further recollections of life on Corfu just before the Second World War – including the excitement and hubbub (and unforeseen comic disasters) accompanying the visit to the town of the returning King of Greece; the invasion of the villa by a Turk, his three wives and a lamb; and the arrival of Adrian Fortescue Smythe, who falls deeply in love with Margo, much to her annoyance. Gerry meets the Rose-beetle Man, from whom he buys three baby Eagle owls, and rescues Hiawatha, a hoopoe, from a hunter. The book comes to an uproarious conclusion with a characteristic Durrell party with a great many guests, each of whom presents a turn in a cabaret.

Durrell's feeling for animals, people and words gives a delightful individual quality to all his writing. As one reviewer has commented: 'Durrell manages to convey not only that he loves animals, but that he enjoys life too – and wants you to enjoy it with him.'

Chapter summaries, textual notes and revision questions

It is a melancholy of mine own ... This quotation on the title-page, from Shakespeare's *As You Like It*, IV,1, is spoken by Jacques, a melancholy lord who is with the banished Duke in the Forest of Arden.

The Speech for the Defence

We are introduced to the family, who insisted on taking up space in the book that was intended for the natural history of Corfu. The author expresses his grateful thanks to Dr Theodore Stephanides, the Durrell family, his wife and his secretary, with a special tribute to his mother. Life on Corfu was like a comic opera.

Alice Through the Looking-Glass Five years after *Alice's Adventures in Wonderland* (1865), Lewis Carroll, whose real name was C. L. Dodgson (1832–98), wrote *Through the Looking-glass*, a sequel.

Corfu Corfu (known in ancient times as Corcyra) is the most northerly of the Ionian islands. It is situated near the coast of Albania, at the mouth of the Adriatic, and together with several other small islands forms a province of Greece. During the fifteenth, sixteenth and seventeenth centuries Corfu belonged to Venice and from 1815 to 1864 it was a British possession. The island is about forty miles long and from fifteen to twenty miles wide. In the north is a mountainous region, full of chasms, caves and underground streams. South of this the island narrows considerably and a few smaller peaks rise from the fertile plains of the eastern side of the island, which produce an abundance of oranges, grapes, honey, wheat and olive oil. The western coast is rocky and dotted with lagoons to the south. Corfu is an island of beautiful scenery and has a summer climate which makes it an ideal holiday resort from March to November. The capital, also called Corfu, stands on a rocky promontory on the east coast, overlooking a fine harbour.

Gerald Durrell's eldest brother, Lawrence, has written an account of his stay there, *Prospero's Cell* (1945), which he describes as 'a guide to the landscape and manners of the island of Corcyra'.

Larry (Lawrence Durrell, born 1912), whose family provided so many distractions from his writing while they were all living on the island, is one of our most distinguished poets and novelists. Lately Director of Public Relations for the Government of Cyprus, he has at various times held the post of Press Attaché in Athens, Cairo, Alexandria and Belgrade, and of Director of British Council Institutes in Greece and Argentina. In 1954 he was elected a Fellow of the Royal Society of Literature. In 1957 he was awarded the Duff Cooper Memorial Prize for *Bitter Lemons*, a book about Cyprus during the troubled years 1953–6, and in 1974 the James Tait Black Memorial Prize for *Monsieur*, the first novel in *The Avignon Quintet*. His *Alexandria Quartet*, consisting of four inter-related novels, is an outstanding achievement. In 1957 he published, for younger readers, *White Eagles over Serbia*, a tale of secret service adventure in the Balkans. His literary achievement to date comprises two volumes of translations from modern Greek writers, two verse plays, seven volumes of his own poetry, more than fifteen novels and several books of travel.

Encyclopaedia Britannica The present edition of this is in 32 volumes, each containing about 1,000 pages.

Dr Theodore Stephanides In *Birds, Beasts and Relatives* Gerald Durrell writes of him: 'Apart from being medically qualified, he was also a biologist, poet, author, translator, astronomer and historian and he found time between those multifarious activities to help run an X-ray laboratory, the only one of its kind, in the town of Corfu.'

This charming person also appears in *Prospero's Cell*, by Lawrence Durrell, which contains the following portrait of him:

> Fine head and golden beard: very Edwardian face – and perfect manners of Edwardian professor. Probably reincarnation of comic professor invented by Edward Lear during his stay in Corcyra. Tremendous shyness and diffidence. Incredibly erudite in everything concerning the island ... Thumbnail portrait of bearded man in boots and cape, with massive bug-hunting apparatus on his back, stalking across country to a delectable pond where his microscopic world of algae and diatoms (the only real world for him) lies waiting to be explored. Theodore is always being arrested as a foreign agent because of the golden beard, strong English accent in Greek, and mysterious array of vessels and swabs and tubes dangling about his person. On his first visit to Kalamai house he had hardly shaken hands when sudden light came into his eye. Taking a conical box from his pocket, he said 'excuse me' with considerable suppressed excitement and advanced to the drawing-room wall to capture a sand-fly, exclaiming as he did so in a small triumphant voice, 'Got it. Four hundred and second.'

split infinitive Separating the *to* and the verb of the infinitive by means of an adverb or an adverb phrase (e.g. 'to always tell the truth') should be avoided wherever possible.

Nirvana It is a belief of Buddhism that by strict spiritual discipline the soul reaches a state of perfect happiness, called Nirvana, where the

individual existence is absorbed into the supreme spirit.
ibis A species of long-legged birds, related to the heron and stork.

Part 1

Dryden John Dryden (1631–1700), poet, playwright and critic was the outstanding man of letters of his age. He became Poet Laureate in 1688. *The Spanish Friar* (1681) was one of Dryden's fourteen plays.

The Migration

Larry suggests that the Durrells should leave Bournemouth and its climate and mentions Corfu. Surprisingly, Mother agrees and soon the family is on its way, each member with a characteristic burden of luggage. Their first sight of Corfu is described.

stertorously With laboured and noisy breathing.
acne A skin disease which causes the face to be covered with pimples.
cleft palate A gap in the roof of the mouth, which makes it difficult to speak.
hag-ridden harassed, trouble (by nightmares or otherwise). A hag was originally a witch.
Rajputana A state of north-western India. Gerry's mother lived for many years in India, where she brought up her young family.
eucalyptus Eucalyptus oil is used as an antiseptic.
iridescence The play of glittering and changing colours.
cicadas Insects with transparent wings. The male cicada makes a shrill chirping sound.

Chapter 1 The Unsuspected Isle

The family and luggage are installed in two cabs, but Roger, the dog, attracts a following of twenty-four other dogs, which detracts from the majesty of their arrival at the Pension Suisse. An endless succession of funerals beneath their balcony window disturbs Mother's peace of mind, and she decides that they must find a house in the country. Mr Beeler, the hotel guide, fails to find a villa with a bathroom. They are continuing the search on their own when they meet Spiro, the taxi-driver, who takes them to a strawberry-pink villa.

Uncle Tom's Cabin This novel by the American writer Mrs Harriet Beecher Stowe was published in 1852. It depicts the sufferings of the

American negroes under slavery, and is said to have done much to hasten the emancipation of the slaves.

Pension Suisse A *pension* is a boarding-house or small hotel on the Continent.

dicky A detachable shirt-front.

filigrees Filigree is a delicate kind of jewel-work made with fine threads of gold and silver.

bathroom Mr Beeler's views on bathrooms are paralleled in this extract from Lawrence Durrell.

> An 'English' house in the island has come to mean a house with a lavatory; and the landlord of such a house will charge almost double the ordinary rent for so remarkable an innovation. Bathrooms are even rarer and are considered a dangerous and rather satanic contrivance. For the peasants a bath is something you are sometimes forced to take by the doctor as a medicinal measure; the idea of cleanliness does not enter into it. Theodore often quotes the old peasant who reverently crossed himself when shown the fine tiled bathroom at the Count's country house and said 'Pray God, my Lord, that you will never need it.' (*Prospero's Cell*)

British Consul An official government representative who resides in a foreign town to protect the interest of British subjects there.

Spiro An abbreviation of Spiridion, the patron saint of the island.

> Spiro is the favourite taxi-driver of the [flying-boat] pilots; they like his Brooklyn drawl, his boasting, his coyness; he combines the air of a chief conspirator with a voice like a bass viol. His devotion to England is so flamboyant that he is known locally as Spiro Americanos. Prodigious drinker of beer, he resembles a cask with legs; coiner of oaths and roaring blasphemies, he adores little children, and never rides out in his battered Dodge without two at least sitting beside him listening to his stories. (*Prospero's Cell*)

Chapter 2 The Strawberry-Pink Villa

The Durrells are immediately attracted by the villa. Spiro takes complete control of their affairs, organizes their move and deals forcefully with a Customs official. Larry installs his many books, but is distracted from his writing by the braying of a donkey and by Leslie's revolver practice. Mother is happily busy in cooking and gardening, and Gerry is fascinated by the creatures that inhabit the garden: spiders, lady-birds, carpenter bees, humming-bird hawk-moths, black ants and cicadas. He discovers how crab-spiders change colour like chameleons, how a certain kind of spider hunts its prey, and how the lacewing fly lays its eggs. He is delighted to discover an earwig's nest. He comes to know the peasant girls who work among the olive trees and in the vineyards; gradually he begins to understand their language and to learn their names, and to identify their homes and relations. The magic of the island settles on the whole family.

bougainvillaea A kind of tropical plant. It has small flowers almost hidden by large magenta or brick-red leaves. Owing to its rank growth it requires plenty of space.

drachmas The drachma is the modern Greek monetary unit; 276 drachmas are worth £1 at present (June 1990).

Christian Science A system of beliefs held by the small group of Christians who accept the theories of the American, Mrs Mary Baker Eddy. The most important of these beliefs is that disease is merely a delusion of the mind; since it has no real existence, disease may be cured without medical treatment, merely by the patient's faith.

Indian Mutiny The great revolt of native troops against British rule in India, 1857–8.

Carpenter bees This kind of bee is dark violet-blue. It is found in Asia, Africa, America and southern Europe, and gets its name from its habit of building its nest in a piece of half-rotten wood.

proboscis A long flexible tube forming part of the mouth of some insects.

chameleon A reptile of the lizard family, which can control the pigment-bearing cells in its skin so as to change colour according to its environment.

Lilliput In the first part of *Gulliver's Travels* by Jonathan Swift (1667–1745) Gulliver is shipwrecked on the island of Lilliput, where the inhabitants are six inches high, with everything else in the island in proportion.

Chapter 3 The Rose-Beetle Man

The day begins with breakfast in the garden, a meal Gerry hurries through in order to begin his explorations, with Roger as his constant companion. He makes friends with a great many of the country people – like Agathi, who sits spinning in the sun and teaches Gerry to sing peasant songs; and Yani, the shepherd who warns him never to sleep under cypress trees. One of the most fascinating characters is the Rose-beetle Man, from whom Gerry buys a tortoise, which he calls Achilles. Achilles loves grapes, strawberries and human company; but he is later found dead in a well and is given a solemn funeral. From the Rose-beetle Man Gerry buys an ugly pigeon, christened Quasimodo by Larry. It is a pigeon with a love of music, which turns out to be female and which, during a lesson with George, upsets a bottle of green ink. One day Gerry and his mother buy the Rose-beetle Man's entire stock of beetles and let them all go free. The last meeting with the Rose-beetle Man is described.

cataract A disease of the eye in which the lens becomes opaque.
hoopoe A bird related to the hornbill. It has a crest of cinnamon-red feathers, tipped with black.
fiesta A local festival, with processions and dances.
English Corfu was a British possession from 1815 to 1864.
Achilles The name of a Greek hero in Homer's *Iliad*. His epithet is 'fleet-footed'.
sweet williams Garden flowers of the species of pink.
Quasimodo A hunchback in Victor Hugo's historical romance, *Notre-Dame de Paris* (1832).
Sousa John Philip Sousa (1854–1932), American bandmaster and composer of stirring marches like 'The Washington Post and 'The Stars and Stripes for Ever'.

Chapter 4 A Bushel of Learning

The family decides that Gerry needs some education and arranges for private tuition with George, a friend of Larry's. George uses a strange assortment of books but encourages Gerry's passion for natural history. While Gerry struggles with mathematics, George practises fencing strokes and the steps of peasant dances. In geography they draw elaborate maps, packed with exciting flora and fauna. History becomes interesting only when it is seasoned with completely imaginary zoological detail. The lessons are attended by Roger, and, on one occasion, by Achilles. Sometimes George and Gerry have outdoor lessons, lying in the shallow water on the beach, and sometimes playing games with 'water-pistol' sea-slugs.

Rabelais François Rabelais (1490–1553), a French satirist. His writings mix wisdom and nonsense, with passages often regarded as gross buffoonery.
eulogistic Full of praise.
Wilde Oscar Wilde (1854–1900) was the author of several witty

comedies, e.g. *The Importance of Being Earnest*.

Gibbon Edward Gibbon (1737–94), a historian. He wrote the famous *Decline and Fall of the Roman Empire*.

Le Petit Larousse An illustrated French dictionary.

saturnine Cold and gloomy. In the Middle Ages the planet Saturn was believed to bestow a sluggish and melancholy temperament upon those who were born under its influence (a belief still held by those interested in astrology).

Herculean Difficult to accomplish. The Greek mythological hero Hercules had to undertake twelve very difficult tasks, known as the Labours of Hercules.

foil A kind of small-sword used in fencing.

tapirs The tapir is an animal which resembles the pig and has a short fleshy trunk. One species is found in South America and another in the Malay peninsula.

Hannibal A Carthaginian general (247–183 BC), who led his army from Spain across the Pyrenees and the Rhone to attack the Roman armies in Italy. He crossed the Alps with 50,000 foot soldiers, 9,000 horses and 37 elephants, and inflicted many severe defeats on the Romans.

Carter Paterson A firm of removal contractors.

hermit crab These crabs inhabit the shells cast off by various sea creatures. The hermit crab carries the shell around with it and as it grows it discards it for another shell of larger size.

Hardy Sir Thomas Masterman Hardy (1769–1839) was Nelson's flag-captain of the *Victory*. Nelson died in his arms at the Battle of Trafalgar in 1805.

Chapter 5 A Treasure of Spiders

One afternoon Gerry and Roger climb to a rocky peak, from which they can see an inviting bay below. They descend to the bay, where they play until hunger drives Gerry in search of food. He rejects the idea of calling on Leonora, Taki, Christaki and Philomena, and decides to visit Yani. Roger chases a cat, and his barking awakens the old shepherd from his siesta. Yani provides Gerry with food and wine, and shows him a scorpion in a bottle of olive oil, explaining how the oil can be used to cure a serpent sting and telling him the story of an old shepherd who died from a scorpion bite. On the way home Gerry rests to eat the grapes that Yani has given him, and notices in a mossy bank a series of tiny trapdoors, which must cover the home of some mysterious creature. Puzzled, he goes in search of George and at

George's villa meets Dr Theodore Stephanides, who explains that the creatures are trapdoor spiders; he visits the bank with Gerry. Gerry is delighted to meet someone who shares his enthusiasm for zoology and is still more pleased when Theodore sends him a pocket microscope and an invitation to tea.

opalescent Glittering with changing colours, like an opal.
blennies Small fish with spiny fins.
virago A fierce warlike woman.
siesta Afternoon rest taken during the hottest hours of the day in Mediterranean and tropical countries.
Gastouri A town situated in the centre of Corfu, south-west of the capital.
Canoni A small town on the coast, south of the capital.
homburg A soft felt hat with narrow brim and dented crown.
daphnia magna A kind of water-flea.
cteniza The name for a family of large burrowing spiders.

Chapter 6   The Sweet Spring

Tea with Theodore becomes a weekly affair, and Gerry learns much from Theodore's conversation, his microscope, and slides and his well-stocked library. Spring returns and the whole island vibrates with life. Larry buys a guitar and works himself (and Mother) into a melancholy mood with Elizabethan love-songs and heavy Greek wine, and suffers from dyspepsia; Margo is obsessed with her personal appearance and a young Turk; Leslie buys a double-barrelled shotgun and takes Gerry with him one morning to shoot turtle-doves.

Sherlock Holmes A famous fictional detective who appears in a number of stories and novels by Conan Doyle (1859–1930).
Darwin Charles Darwin (1809–92), the naturalist whose theory of evolution has had an important influence on modern science and modern thought generally. His best-known work is *The Origin of Species* (1859).
Le Fanu Joseph Sheridan Le Fanu (1814–73), though trained as a lawyer, became proprietor of the *Dublin University Magazine* and earned fame by his twelve novels, noted for their atmosphere of mystery and the supernatural. His most popular novel is **Uncle Silas** (1864).
Fabre Jean Henri Fabre (1823–1915), a famous French naturalist who became known as 'the Insect Man'.
spinnerets The organs by means of which silkworms and spiders spin their threads.

cyclops A kind of small freshwater shellfish which has only one eye in the centre of its forehead. It is named after the legendary one-eyed giants of Greek mythology.

Hephaestus The Greek god of fire and the patron of metal-workers. The Romans called him Vulcan.

Salonika (or Thessalonika) A region of north-eastern Greece.

vampires A species of bat which live by sucking the blood of other animals. (Also the name for an imaginary being – often supposed to be a reanimated corpse – which sucks the blood of sleeping persons.)

Bosnia A region of central Europe, formerly part of the Austrian Empire.

Val de Ropa One of the central plains of the island of Corfu, forming an enclosed basin of soft, sandy soil. The land is so flat that water flows with difficulty, so that considerable irrigation schemes have been carried out.

Vetch A name applied to a wide variety of plants of the bean family.

asphodel A plant with beautiful lily-like flowers, usually white or yellow.

ungulate A hoofed animal.

pinking A car engine pinks when the valves make a high-pitched metallic popping noise.

chairete, kyrioi Be happy, masters (Greek).

Conversation

Larry invites several of his friends to stay and suggests that in order to accommodate them the family should move to a larger villa, but Mother says firmly that they are not moving.

Revision questions on Part 1

1 Summarize what you have learned so far about the personality and peculiarities of each member of the family.

2 Write an account of Gerry's meetings with Yani.

3 Tell the story of the life and death of Achilles.

4 Describe Gerry's lessons with George.

5 Describe the characteristics and habits of the trapdoor spider.

6 What does Gerry learn from Theodore?

Part 2
Chapter 7 The Daffodil-Yellow Villa

Spiro organizes the family's move to the sadly decaying villa, with its derelict garden and orchard. Mother engages as a servant in the house the gardener's wife, Lugaretzia, who loves to discuss her ailments. Tables, chairs and wardrobe fall to pieces and Mother, Margo and Gerry go to buy new furniture in the town of Corfu. Here they are swept along in a vast crowd which is celebrating the local saint's day, and are forced to enter the cathedral in a procession of people who line up to kiss the feet of the mummified body of Saint Spiridion and to request his help. To her mother's horror Margo kisses the saint's feet, hoping that he might cure her acne, and then spends three weeks in bed with influenza, under the care of Dr Androuchelli.

While Margo is being nursed back to health, Larry is engaged in fitting up his library; Leslie converts the veranda into a shooting-gallery; Mother is busy in the kitchen; and Gerry, without a tutor now, explores the fifteen acres of garden with Roger. He finds many old insect friends; and there are swallows too, residing under the eaves of the villa – Gerry watches with particular interest the odd nesting antics of two pairs of swallows. He finds an extraordinary oil beetle; Theodore tells him about its life history and recalls an experience involving a horse.

Albania The island of Corfu is only about ten miles off the coast of Albania, a country north of Greece in the Balkan peninsula.
lugubrious Mournful, sorrowful.
hypochondriacs People affected by depression and melancholy for no real cause.
kyria Master (Greek).

Chapter 8 The Tortoise Hills

In the hills behind the villa, Gerry observes the habits of ant-lion larvae, hunting wasps, caterpillars, mantids, chaffinches and goldcrests. Soon the hills are full of tortoises, emerging from the soil after their hibernation. Gerry watches the males fighting over the females and is fascinated by their clumsy courtships and fumbling matings. He becomes familiar with a one-eyed tortoise (Madame Cyclops) and watches her laying her eggs. Throughout the spring and early summer the villa is filled with a succession of

Larry's friends – each more eccentric than the one before: Zatopec, an Armenian poet; Jonquil, Durant and Michael, three unproductive artists; and Melanie, the hairless Countess de Torro. There is an account of their strange conversations at an extraordinary dinner party one evening and Theodore tells another of his true but fantastic-sounding stories about Corfu.

mantids Insects with long straight wings: they are extremely fierce, killing other insects and cutting them to pieces.
erysipelas A distressing skin infection, which causes the skin to go deep red. Due to the countess's ill-fitting false teeth, Mother misunderstands her and thinks the countess has been suffering from syphilis, a venereal disease.
Lawrence Michael is talking of D. H. Lawrence (1885–1930), poet and novelist. Many of his novels are concerned with basic relationships between men and women. The most extreme example is *Lady Chatterley's Lover* and the publication of the unexpurgated version in 1960 caused a great sensation. Margo's comment shows that she is confused.
The Seven Pillars of Wisdom First published 1926, in a limited edition; by T. E. Lawrence (1888–1935) – the famous 'Lawrence of Arabia'. In World War I, Lawrence, an army officer, was sent from Egypt to help the Arabs in their revolt against the Turks; he was accepted by the Arabs, who treated him as one of themselves. Wearing Arab dress, he entered Damascus with the leading Arab forces in 1918. The book deals with Lawrence's experiences in this desert war.
shibboleths The old-fashioned and generally abandoned doctrines of particular parties or sects. (From the Hebrew: see Judges xii,6).

Chapter 9 The World in a Wall

Gerry finds great interest in the crumbling wall surrounding the sunken garden. There is a whole miniature landscape of varied vegetation and it houses a mixed lot of creatures – toads, geckos, crane-flies, moths and beetles by night; and wasps, caterpillars and spiders by day. Gerry grows very fond of scorpions and is able to observe their courtship dances. He finds a female scorpion with a mass of babies clinging to her back, but Larry opens the matchbox in which they have been placed and scatters them over the dining-table. The results are immediate panic, and the eventual decision that Gerry must receive more education. He is taught by the Belgian consul, who lives in the Jewish quarter of the town and continually rushes to the window to shoot cats.

Mother's embarrassing encounters with the consul are described. On Thursdays, Theodore visits the villa, enjoys watching the seaplanes land and goes off collecting specimens with Gerry. Theodore tells a story of a badly planned building.

geckos The gecko is a kind of lizard with a rather triangular-shaped head and sucking-pads on its toes, so that it can run up a wall and even across a ceiling. It is sometimes called a house-lizard.
predators Animals that prey on other animals.
antimacassars Covers for the backs of sofas and chairs to protect them from hair-grease. The name dates from the middle of the nineteenth century, when Macassar oil, made from ingredients that came from Macassar in the East Indies, was a popular hair-grease.
rickets A disease caused by a deficiency of vitamin D. The composition of the bone is changed; the most noticeable symptom is that the sufferer's legs are bowed outwards.

Chapter 10 The Pageant of Fireflies

Summer arrives and the island vibrates with the noises and movements of insects. There is a new tutor, Peter, a young man fresh from Oxford, whose strict ideas on education gradually relax. Gerry begins to write a book, and Peter shows great interest in Margo. Gerry has a large room of his own in which he houses his creatures; one of these, a stuffed bat, causes some concern. Searching for another bat, Gerry studies a number of other night creatures, and one day he captures a young owl, which he calls Ulysses. Roger and Ulysses are introduced to each other. Ulysses goes out hunting every night and sometimes accompanies Gerry and Roger for a late evening swim. The family takes to bathing at night, and on one occasion Gerry encounters a group of porpoises. Mother buys a bathing-costume, much to the family's amusement, and they celebrate her first entry into the sea with a moonlight picnic, a highly amusing occasion. During the hot months the sea becomes phosphorescent, and the family witness a fantastic spectacle of porpoises, fireflies and phosphorescence together.

Fireflies This is the name popularly given to a whole group of creatures that give out light, though in fact most of them are beetles, not flies; the glow-worm is one of these. Fireflies found in warmer countries give out such a brilliant light that one can read and write by it.

zithered Made a noise like the sound of a zither, an Austrian musical instrument with strings fixed to the rim of a shallow box.

quintessence The perfect embodiment, the purest form of some quality. In ancient and medieval times, when all material things were held to be composed of four elements, the quintessence ('fifth essence') was supposed to be the substance of which the heavenly bodies were composed and to be latent in all things.

Boy's Own Paper This famous weekly magazine was founded in 1879. It was eventually published monthly, before publication ceased completely. It was well known for its tough adventure stories, and among its contributors were Conan Doyle, Jules Verne, Talbot Baines Reed, R. M. Ballantyne and G. A. Henty.

taxidermy The art of preserving the skins of animals and stuffing and mounting them so that they have the appearance of living animals.

Scops owl A small horn-owl (so called because of the long horn-like tufts of feathers on the head).

Ulysses The Roman name for Odysseus, the Greek hero in Homer's *Odyssey*. There are legends that it was on the island of Corfu that Ulysses met Nausicaa, a princess who looked after him after he was shipwrecked.

Maltese crosses In a Maltese cross each arm broadens out from the centre and has an indentation at its extremity. This is the form of cross used as the symbol of the St John Ambulance Brigade.

emaciated Lean, with wasted flesh.

Milky Way A galaxy of stars, which appears as a luminous band across the night sky.

phosphorescence This phenomenon, by which the sea sometimes becomes luminous at night, is the result of the emission of light from the bodies of marine organisms. The same kind of luminousness appears in glow-worms and fireflies.

Albert Memorial A large memorial erected in Kensington Gardens, London, in memory of Prince Albert, consort of Queen Victoria.

Perseus's rescue of Andromeda In Greek mythology Andromeda was the daughter of Cassiopea, Queen of Ethiopia. The latter angered Poseidon, god of the sea, by boasting of her daughter's beauty, and Poseidon sent a sea-monster to attack the country. In order to abate the monster's wrath, Andromeda was exposed on a rock. She was rescued by the hero Perseus, who had just conquered the Gorgons and used the head of Medusa to change the monster into a rock. (See note on 'a Medusa head', p.25).

Chapter 10 The Enchanted Archipelago

Gerry discovers a group of small islands full of attractive sea fauna and wishes for a boat of his own. He compiles a list of birthday presents that he would like, and succeeds in getting Leslie to promise to build him a boat. While Leslie is doing this, Gerry makes himself ponds for his sea-creatures. The boat is completed, approved and named *The Bottle-Bumtrinket*. At its launching it turns turtle, throwing Peter overboard. Spiro helps with the preparations for a birthday party, which is attended by scores of guests and which concludes with the dancing of the *Kalamatiano*. Next day Gerry sets off early in his boat to explore the islands, observes the alarm signals of clams and serpulas, and studies many kinds of sea-creatures – blennies, sea-urchins, anemones, crabs and a baby octopus. He returns home with a large collection of specimens.

Archipelago A stretch of sea dotted with many islands.
formalin A pungent-smelling liquid, used by biologists as an antiseptic for preserving zoological specimens.
ballistic Concerning the projection of missiles.
Kalamatiano A national dance of Greece, thousands of years old, which takes its name from Kalamata, an ancient town in southern Greece.
clams Shellfish of the same family as oysters, mussels and cockles, with shells consisting of two halves hinged together.
serpulas A serpula is a kind of worm which builds itself a 'shell', a crooked tube made of lime, which it attaches to rocks in the sea.
a Medusa head In Greek mythology Medusa was one of the Gorgons, three sisters with serpent-entwined hair, hands of brass and bodies covered with scales. They turned to stone anyone who gazed upon them. With the aid of a helmet from Pluto and a shining shield from Athene, which he used as a mirror, the hero Perseus cut off Medusa's head.
sea-fans A class of marine animals that includes corals and sea anemones; this particular creature has a shape that branches out like a fan.

Chapter 12 The Woodcock Winter

Peter's services as tutor are dispensed with and Margo is broken-hearted. Leslie's shotgun burglar alarm causes panic one night. Margo takes Gerry's boat in order to be alone on an island. However, a sirocco is raging, and Margo returns only with great difficulty. The signs of approaching winter are described, and the shooting season arrives, much to the delight of Leslie, who goes boar-hunting. Larry makes scathing comments and is challenged by Leslie to join him in shooting birds next morning. They go down to the swamp, and Larry, shooting at two snipe, falls into a ditch and, after much effort, is rescued. He retires to bed with a bottle of brandy and a huge fire, which sets fire to a beam under the floor-boards and causes another panic, during which Larry directs operations from his bed.

Woodcock A game-bird closely related to the snipe and much esteemed as food.
Timber The lumberjack's warning cry when a tree he is felling is about to fall.
siroccos Hot winds blowing from the north coast of Africa over the Mediterranean.
snipe A bird found in almost every part of the world and remarkable for the length of its bill.
mercurial Quick-witted. In the Middle Ages the planet Mercury, with the other planets, was believed to bestow special qualities of temperament on those who were born under its influence. cf. 'saturnine' (see note p.oo) and 'jovial' (from Jove or Jupiter).
Shelley The poet, Percy Bysshe Shelley (1792–1822). He and a friend were drowned while sailing their small yacht off the north-west coast of Italy.
coup de grâce Finishing stroke (Fr). Originally it was the blow by which a person who had been mortally wounded was put out of his misery.
full fathoms five cf. Ariel's song in Shakespeare's *Tempest*, I,2. 'Full fathom five thy father lies; Of his bones are coral made.'
holocaust Complete destruction by fire. Originally the word was used for a sacrifice wholly consumed by fire.
keep your head . . . losing theirs cf. Rudyard Kipling's poem, *If*. 'If you can keep your head when all about you/Are losing theirs and blaming it on you.'

Conversation

Spring returns. Great-aunt Hermione wishes to come and stay with the family. To prevent this, Larry suggests that they move to a smaller villa.

a change is as good as a feast Margo confuses two proverbs: A change is as good as a rest, Enough is as good as a feast.

Revision questions on Part 2

1 Write a character-sketch of Lugaretzia.

2 What do you know about Saint Spiridion?

3 What happens to Mother, Margo and Gerry in the cathedral?

4 Describe Gerry's discovery of the tortoises.

5 Write an account of the incident which involves the female scorpion and her babies.

6 Describe Gerry's lessons with the Belgian consul.

7 How does Gerry come to find Ulysses?

8 Write an account of the moonlight picnic.

9 Write an account of the creatures that Gerry sees when he explores the islands in his boat.

10 Tell the story of Larry's hunt for snipe.

Part 3

UDALL Nicholas Udall (1505–56) was Headmaster of Eton and later of Westminster School. He had a considerable reputation as a scholar and a writer of Latin plays. *Ralph Roister Doister*, which he wrote for the St Andrew's Day celebrations at Eton (probably in 1553), is the earliest known English comedy.

Chapter 13 The Snow-White Villa

Spiro finds the family a new villa. Gerry learns more about mantids, which wage a constant war against the geckos, and describes the habits of a particularly ferocious gecko, christened Geronimo, who has a quick method of dealing with lacewing flies, moths and rival geckos. Interested in the breeding habits of mantids, Gerry takes home a large mantis which he believes is pregnant and which he christens Cicely. She is attacked by Geronimo and there follows a fierce and prolonged battle, which is ultimately fatal for Cicely. Gerry comes upon two giant toads, which are received without enthusiasm by Spiro and the rest of the family, but with great interest by Theodore, who feeds one of them with a large worm.

Geronimo The leader of a North American Apache Indian band who showed great courage and cunning in raids on the Mexicans, who had killed his family. Massacres in Arizona and New Mexico, however, brought him up against the US army. Time after time Geronimo managed to escape capture, eluding five thousand troops with his band of thirty-five men. Eventually he was taken and confined on a farm, which he was allowed to work. He died in 1909.
viscera The internal organs of the body.
carunculated Having small outgrowths of flesh.
Buddhas Statues of the founder of Buddhism usually depict him as a rather rotund figure in a sitting position.
flaccid Limp.
to throws i.e. to throw up, to vomit.
maw Stomach.
toad *in a hole* Theodore is punning on the name of a dish consisting of sausages baked in batter.

Chapter 14 The Talking Flowers

Another tutor is found: Kralefsky, an eccentric bird-lover. He and Gerry spend much time feeding Kralefsky's birds, picking groundsel in the garden and taking walks which invariably lead them to the bird market. But Kralefsky is a stickler for work, though Gerry finds his lessons very dull – even French, which is taught from a set of bird books. One morning Gerry meets Kralefsky's mother, a tiny old lady with long hair, who lies in bed in a room full of flowers and talks to Gerry about her hair, about flowers holding conversations and about a rose being harried to death by Michaelmas daisies.

black redstart The redstart is a bird of the thrush family and is found in almost all parts of Britain. It has a red tail (*steort* is the Old English word for 'tail'), a white forehead, a black throat and a reddish-brown breast. The black redstart has a sooty-black breast and visits Britain only occasionally.
That's the ticket i.e. that's what is wanted; that's the correct thing.
aviculturist Bird fancier or bird breeder.
Alice in the Looking-glass garden In the second chapter of Lewis Carroll's *Through the Looking-glass* Alice finds that all the paths in the Garden of Live Flowers eventually lead her back to the house.
Narcissus In Greek mythology Narcissus was a beautiful youth who fell in love with his own reflection in a fountain, and died of despair because he could not reach the object of his love.
Touched 'Cracked', slightly simple.

Chapter 15 The Cyclamen Woods

The Cyclamen Woods are three tiny olive-groves on a hill near the villa. Here Gerry finds a magpie's nest, from which he takes two babies. Larry is convinced that they have criminal instincts; Spiro christens them 'The Magenpies'; and they are allowed to join the family on condition that they do not steal. One day they get into Larry's room and cause havoc; Larry is furiously angry and Gerry decides that the Magenpies must be kept in a large cage. Gerry invites Kralefsky to spend a day at the villa, so that he can help to build the cage and also teach him how to wrestle. Kralefsky has told Gerry many anecdotes of imaginary adventures in which he rescued ladies from bull-terriers in Hyde Park, icebergs near Murmansk, bandits in the Syrian desert, a firing-squad behind the enemy lines and a brute of a man in Paris who turned out to be the champion wrestler of France. Gerry persuades him to demonstrate some wrestling holds and in the course of practice throws Kralefsky to the floor and cracks two of his ribs, so that, to the accompaniment of unhelpful comments from Larry, Kralefsky has to be driven to Theodore for an X-ray.

Arsène Lupin The hero of a series of crime novels by Maurice Leblanc (1864–1941).
anthropomorphic Attributing human personality to anything impersonal.
Attila the Hun The Huns were a warlike race of nomads who invaded Europe about the year 375. Later, under their leader Attila, they overran a great part of the Roman Empire before they were defeated in 451. Attila's cruelty earned him the name of 'the scourge of God'.
Hyde Park A large park in London.
Murmansk A port in the Kola Peninsula, on the northern coast of the USSR.
Neanderthal Man The race of men living in the Palaeolithic or Old Stone Age; so called because parts of a human skeleton from these times were found in a cave in Neanderthal in Germany in 1857. Neanderthal man would seem to have had a large head, with a sloping forehead and a heavy ridge above the eyes, a short but powerful body, covered with hair and carried in a slouch on short legs.
Anopheles One of several kinds of mosquito; the type that carries the disease of malaria through parasites which live in its blood and which are transferred to human beings by the mosquito's bite.

Chapter 16 The Lake of Lilies

Confined in their cage, the Magenpies drive the family, the dogs and the hens frantic with their imitations of human voices. The three dogs are now joined by Dodo, a Dandy Dinmont bitch with limited intelligence and a hind leg that frequently comes out of joint. She causes problems when she comes into season and (thanks to Puke) has a puppy. In the season of lilies the family, the dogs, Spiro, Theodore and a maid, travelling with some difficulty in both the *Sea Cow* and the *Bottle-Bumtrinket*, visit a large lake separated from the sea by a wide sand-dune. Larry and Margo sleep; Mother goes collecting plants; Spiro catches fish; Leslie goes off shooting on one side of the lake, while Gerry and Theodore collect specimens on the other side. They all assemble on the beach for a lunch cooked by Spiro and then sleep through the heat of the afternoon until teatime. The appearance of a robin reminds Theodore of an opera singer, and he tells them of the last opera they had in Corfu and of the unfortunate experience of the singer who had to throw herself from the battlements of a fortress. After tea Gerry and Theodore continue their investigations at the edge of the lake, until Spiro cooks supper and the family sails home in the moonlight.

slicker A sophisticated city-dweller.
Frankenstein Monster. Strictly, Frankenstein is the hero, a student of natural philosophy, of the novel of the same name by Mary Shelley, the poet's second wife. Frankenstein forms, from human remains, an animate creature, man-like but of revolting appearance. By confusion, the name of the creator is today often applied to his creation.
dryads Wood nymphs in Greek mythology.
Tosca A well-known opera by the Italian composer Puccini (1858–1924).
accolade This is the technical name for the salutation (an embrace, a kiss, or a stroke on the shoulder with the flat of a sword) which marks the bestoval of a knighthood. Here it means favour, distinction, honour.

Chapter 17 The Chessboard Fields

Between the villa and the sea is an area of fields, each one edged with narrow waterways. Here Gerry finds a multitude of creatures, and on the flat sands beyond are many sea-birds. He decides to try once more to catch Old Plop, an ancient terrapin,

Chapter 17 The Chessboards Fields

but the latter is disturbed by the dogs, who go chasing a lizard and come upon two water-snakes lying in the sun. Gerry catches one of these, but the other escapes into the mud. While he is capturing the second one he is watched by a man who turns out to be a convict, allowed out of prison on weekend leave.

The convict takes him to share some cockles in his boat, where Gerry finds Alecko, an immense black-backed gull, a fierce bird that is nevertheless friendly towards Gerry. The convict says that Gerry can have this magnificent bird and that they may meet the following morning to catch some fish for Alecko. Gerry learns that the convict's name is Kosti Panopoulos and that he killed his wife. With some difficulty Gerry gets the bird home. Larry is convinced that it is an albatross and that it will bring ill-luck to the family. Mother is shocked to learn that Gerry was given the gull by a convict and a murderer, but eventually allows Gerry to go fishing with him and even to bring him to the villa.

Venetian days The island of Corfu belonged to Venice from the fifteenth to the seventeenth century.

salt pans Natural or artificial basins, where water collects and leaves a deposit of salt on evaporation. Large quantities of salt are still produced in Corfu.

terrapins The popular name for several species of turtles and tortoises.

pipe-fish A species of fish with a long tapering body and jaws which form a tube or pipe, with the mouth at the tip.

oyster-catcher This bird is popularly known as the sea-pie. It has black and white plumage and a long, wedge-shaped orange bill. It lives on the coast and feeds on shellfish.

dunlin Sometimes known as the red-backed sandpiper, this bird too has a long bill. It lives on sandy, muddy shores, breeds in the arctic regions and winters around the Mediterranean.

tern Another kind of sea-bird, smaller and more slender than a gull.

carapace Literally, the shell of a tortoise.

albatross This is the largest sea-bird known. Some specimens have measured 17½ feet from wingtip to wingtip. Albatrosses sometimes accompany ships for long periods, and because of this sailors believe that it is unlucky to kill an albatross.

Roc A legendary bird of great size and strength.

Ancient Mariner *The Rime of the Ancient Mariner* by S. T. Coleridge (1772–1834) tells how the mariner's shooting of an albatross brings a curse on the ship, which is becalmed, and on the crew, who die of thirst. The spell is finally broken, but the mariner is condemned to wander the seas for ever.

guano The excrement of sea-fowl, used as manure.

Saint Elmo's fire A phenomenon that sometimes produces in stormy weather the appearance of flames on the masthead and yardarms of a ship (or on the wingtips of an aeroplane). It is caused by a discharge of electricity. St Elmo was a third-century Italian bishop who became the patron saint of sailors.

Chapter 18 An Entertainment with Animals

The family decides to give a Christmas party (in September!) and the house is alive with the activity of preparation. The party is almost completely taken over by the animals. In reorganizing the terrapin pond to provide accommodation for Old Plop, Gerry decides that he requires goldfish; Spiro and a friend eventually steal some for him from the garden of a palace. On the morning of the party Mother finds that Dodo has come into season and is being followed into the kitchen by canine Romeos, while Gerry discovers that two of the goldfish have been eaten, so he has to rearrange the reptiles in the pond. Owing to the heat of the sun the watersnakes are put in the bath. The Magenpies escape from their cage and, drunk with beer, cause havoc to the table that has been laid for lunch. Alecko is also missing.

While the guests are talking together, Leslie returns from hunting and soon afterwards emerges from the bathroom, clad in a small towel and complaining loudly of snakes. The subsequent heated discussion eventually subsides as the guests sit down to lunch. The next moment there is an uproar when some of them are bitten by Alecko, who is under the table. After much manoeuvring Gerry succeeds in taking Alecko back to his cage. Kralefsky relates how a friend of his rescued a lady from an attack by a large gull, and Larry tells the story of an eccentric aunt who was almost killed by bee-stings and a large iron bucket, and who set fire to her cottage while trying to smoke out a swarm of bees. Theodore follows this up with some funny stories about the Corfu Fire Brigade. The guests consume huge quantities of food, and spend the day talking and swimming. Spiro brings three turkeys cooked by his wife. Dodo rushes into the crowded drawing-room, pursued by a pack of belligerent dogs. The latter are set upon by Roger, Puke and Widdle. In the attempts to quell the fighting Spiro drenches the dogs (and some of the guests) with water. Everyone settles down and the party comes to

Chapter 18 An Entertainment with Animals

a successful conclusion against the background of an enchanted moonlit countryside.

milk of human kindness cf. Shakespeare's *Macbeth*, I,5, where Lady Macbeth says of her husband, 'Yet I do fear thy nature;/It is too full o' the milk of human kindness.'

switched the points A metaphor from points on a railway, which direct a train from one line to another.

Lamb Charles Lamb (1775–1834) was the author of *Essays of Elia*, several poems, and some criticisms of Elizabethan dramatists. With his sister Mary he wrote the famous *Tales from Shakespeare*.

minotaur This fabulous monster, half bull and half man, was fed on human flesh. It was kept in the centre of a maze by Minos, a legendary king of Crete, who demanded from the conquered Athenians an annual tribute of youths and maidens. Eventually Theseus, son of the Greek king, killed the minotaur with the help of Ariadne, Mino's daughter.

hamadryads Another name for king cobras, large and very poisonous snakes found in India and China.

Saint Francis of Assisi As a young man Francis, who was a member of a wealthy Italian family, lived a life of pleasure, but after a serious illness he suddenly turned to a life of devotion, solitude and prayer, giving up all his possessions in order to look after the poor. He was joined by disciples and he formed them into a new order of friars, the Franciscans, often known in England as Greyfriars. He laid special stress on helping the sick, and he is particularly associated with a love of nature and of animals (especially birds). He died in 1228 and was canonized two years later.

beasts in his belfry The usual phrase, is, of course, '*bats* in the belfry'. cf. 'batty', crazy.

all the nice gulls 'All the nice *girls* love a sailor' was a popular song earlier in this century.

gull and wormwood 'Wormwood and gall' is a Biblical phrase for what is bitter and vexing. See Lamentatations iii,19: 'Remembering mine affliction and my misery, the wormwood and the gall.' Gall, the secretion of the liver (or bile), typifies an intensely bitter substance. Wormwood is a plant (*Artemisia absinthium*) with a very bitter taste. It is used in making the liqueurs, absinthe and vermouth (the latter being the German word for wormwood).

Punch This weekly illustrated journal of literary and pictorial humour was founded in 1841 and still continues to be published.

fetish In its original sense a fetish is an inanimate object worshipped by savages, who believes it to possess magical powers. In a wider sense (as here) the word is applied to anything regarded with unreasonable reverence or esteem.

Talking of fires In *Prospero's Cell* Lawrence Durrell gives an account by Spiro of another funny incident concerning the Corfu Fire Brigade.
gastronomic Relating to the art and science of good eating.

The return

Kralefsky can teach Gerry no more, and Mother decides that the family must return to England for a time. Everything is packed and goodbyes are exchanged with all their peasant friends. Spiro deals with the Customs man. Theodore, Kralefsky and Spiro see the family and all the animals on to the boat. At the Swiss frontier a passport official describes the family as: One travelling Circus and Staff.

Revision questions on Part 3

1 Describe the activities of Geronimo.

2 Describe Gerry's visit to Kralefsky.

3 Write an account of his meeting with Mrs Kralefsky.

4 Give an account of the day spent by the family at the lake of lilies.

5 Describe Gerry's encounters with the convict.

6 Write an account of the unfortunate happenings at the Christmas party.

7 In your own words tell one of Theodore's stories about the Corfu Fire Brigade.

Characters and creatures

Since the book contains so many characters and creatures, some of whom are only mentioned briefly from time to time, the student will find it useful to have a list of these, with chapter and page references to the first appearances of the less important ones.

My family

Mother, Larry, Leslie, Margo, Gerry and Roger the dog. At the start of their five-year stay on the island of Corfu Larry is twenty-three years old, Leslie is nineteen, Margo is eighteen and Gerry is ten.

Other characters

Mr Beeler, the hotel guide (Chapter 1, page 26); Spiro (2,28); the Rose-beetle Man (3,46); George, a tutor (4,57); Dr Theodore Stephanides (5,74); Lugaretzia, the gardener's wife (7,100); Dr Androuchelli (7, 105); the Belgian consul, another tutor (9,132); Peter, another tutor (10,142); Kralefsky, another tutor (14,214); Mrs Kralefsky, his mother (14,223); Kosti Panopoulos, a convict (17,266).

Peasants on Corfu

Agathi (3,43); Yani (3,44); Leonora (5,67); Taki (5,67); Christaki (5,67); Philomena (5,68); Aphrodite, Yani's wife (5,70).

Other Animals

Achilles, the tortoise (3,48); Quasimodo, the pigeon (3,52); Madame Cyclops (8,117); Ulysses, the Scops owl (10,145); Widdle and Puke, the dogs (11,165); Geronimo, the gecko (13,200); Cicely, the mantis (13,203); the Magenpies (15,232); Dodo, the Dandy Dinmont (16,246); Old Plops, the terrapin (17,262); Alecko, the black-backed gull (17,268).

Notes on the more important of these appear at the end of this section.

There are many other creatures, which are not given personal names, but which are described at some length: spiders (2,36); earwigs (2,38); trapdoor spiders (5,73); swallows (7,106); oil-beetles (7,109); tortoises (8,112); geckos (9,127); scorpions (9,128); mantids (10,141); porpoises (10,150); various sea creatures (11,168–70); toads (14,209); Kralefsky's birds (15,216); water-snakes (17,264); goldfish (18,279).

It would be a mistake to approach the human characters of *My Family and Other Animals* in the same way that one would approach the characters in a novel. They are not imaginative creations, like the characters of Dickens, Scott or Jane Austen, and the author does not present them as such. They are incidental to his main purpose, which is to give an account of the five years of his childhood which were spent on Corfu. Durrell says that he has 'attempted to draw an accurate and unexaggerated picture' of his family. It is perhaps doubtful whether some members of the family (Larry, for example) would agree that he has been wholly successful in that attempt! He tends to dwell on the more eccentric aspects of his characters; this is very noticeable, for example, in the case of Leslie, who appears as little more than a young man with a gun. None of the characters is presented with any suggestion of development: we leave them exactly as we found them. We see nothing of Larry's progress *as a writer*, for instance.

The author would no doubt reply that he is not a novelist, that he is not concerned with the creation and development of character, and that in any case the members of his family *were* rather eccentric. 'They appear,' he says, 'as I saw them.' It is perhaps only to be expected that they should be a little one-sided and somewhat caricatured, for they are the inhabitants of a child's world and a world, moreover, whose foreground is occupied by animals. But it is nevertheless a world on which he looks back with adult insight and adult imagination, and the adult Durrell appears to feel the need for a word of explanation for his portrayal of the family.

To explain some of their more curious ways, however, I feel that I should state that at the time we were in Corfu the family were all quite young: Larry, the eldest, was twenty-three; Leslie was nineteen; Margo eighteen; while I was the youngest, being of the tender and impressionable age of ten' (p.9).

Gerry

'that boy's a menace . . . he's got beasts in his belfry.'

Though his mature self is the narrator of the book, Gerry must surely rank as its chief character. He is a friendly, inquisitive, outgoing boy, but even at the age of ten he is already in the grip of his lifelong obsession with animals – all animals. This is revealed at once in the book, in 'The Migration', p.18:

I travelled with only those items that I thought necessary to relieve the tedium of a long journey: four books on natural history, a butterfly net, a dog, and a jam-jar full of caterpillars all in imminent danger of turning into chrysalids.

Gerry is given no lines of actual dialogue in the book – but he is the cause of much dialogue in others – the villagers, his many tutors and, above all, his family – affectionate but often infuriated by the antics of Gerry's large collection of animal friends and pets, many of which share his bedroom but also roam at will throughout the house and garden.

Like Mother he is of a placid disposition, but earns his own place in a family that seems – along with friends and distant relatives, as well as employees – to be made up entirely of eccentrics. He seldom, if ever, loses his temper, but sometimes inspires near-frenzy in his more excitable siblings – Larry in particular, who directs some of his funniest exaggerations at (or about) Gerry and his animal associates.

There is, for instance, the occasion when, having housed his family of scorpions in a matchbox, he leaves it on the drawing-room mantelpiece; Leslie, about to light a cigarette, picks it up and brings it into the dining-room. Soon the entire scorpion family is scattered down the length of the dining-table, and Larry shouts, 'It's that bloody boy . . . he'll kill the lot of us . . . look at the table . . . knee-deep in scorpions . . . Every matchbox in the house is a deathtrap' (2,9,131). Or when Gerry's magpies make a shambles of Larry's bedroom: '"I did not ask for a lecture on the crow tribe," said Larry ominously . . . "I am just telling you that you will have to get rid of them or lock them up, otherwise I shall tear them wing from wing"' (3,15,235). And, again in the dining-room, when Alecko, Gerry's black-backed gull (which Larry insists on calling an albatross) gets under the table at lunchtime: '"Don't move . . . keep quite still, unless you want your legs taken off at the knee!" Larry informed the company' (3,18,288).

Maddening small boy that he must have been at times, Gerry obviously is affectionately tolerated by everyone (including his tutors, who all treat him as their equal in age): this must have been the result of his own friendly nature and his obviously genuine dedication to natural history.

Mother

Like a gentle, enthusiastic and understanding Noah, she steered her vessel full of strange progeny through the stormy seas of life with great skill.

There is no doubt that Mother must have had many an anxious moment on account of her extraordinary family – a family that persuades her to move house at the slightest opportunity; holds a Christmas party in September and invites forty-five guests instead of ten; and comprises individuals of such different and definite personalities, each bent on expressing itself to the utmost. It is not surprising that Mother frequently appears rather vague, can never remember the date of her birth, and is often not quite sure what is happening, preoccupied as she is with her passion for gardening or cooking, 'spectacles askew, muttering to herself'.

She is handicapped by her limited knowledge of Greek and terrified if she has to speak with the Belgian consul. But she always seems to retain her dignity, even when Larry has persuaded her to drink too much wine. She remains calm and imperturbable, continually pacifying her rebellious brood. She understands her family well, is guarded when they make their impossible suggestions, and attempts to be firm; but every member of the family seems to get what he wants in the end. This is because she has an innate kindness towards everyone, people *and* animals.

Even the Magenpies, whose behaviour must have been infuriating, are not to be blamed: they cannot be held responsible, she says, if they are drunk. Whenever Gerry brings new animals home her first thoughts are about feeding them; and if the water-snakes can be saved only by a long, cool bath – well, then, into the bath they must go. It is true that logic is not her strong point. She is superstitious about peacock feathers and the cook that died, and she thinks that Leslie, when taking burglars by surprise, should ring a bell before shooting at them, in order to warn the family. But she shows extraordinary patience and sympathy: with so many animals and so many strong-minded people to contend with, she is indeed 'a gentle, enthusiastic and understanding Noah'.

Larry

Larry was designed by Providence to go through life like a small blond firework, exploding ideas in other people's minds, and then curling up with cat-like unctuousness and refusing to take any blame for the consequences.

Larry, it is clear, sees himself as first mate of a troublesome crew with an unreliable captain. He is always making plans to organize the family and is hurt when they do not immediately agree with them. It is his idea to go to Corfu in the first place, his idea to change villas *twice*. Organizing is easy: it is merely a matter of applying intelligence.

Larry was always full of ideas about things of which he had no experience. He advised me on the best way to study nature, Margo on clothes, Mother on how to manage the family and pay her overdraft, and Leslie on shooting. He was perfectly safe, for he knew that none of us could retaliate by telling him the best way to write. Invariably, if any member of the family had a problem, Larry knew the best way to solve it; if anyone boasted of an achievement, Larry could never see what the fuss was about – the thing was perfectly easy to do, providing one used one's brain.

Moreover, he is quite irrepressible. After scorning Leslie's shooting achievements and being challenged, he wriggles a little but accepts, and we next see him hunting snipe, 'a small, portly and immensely dignified Robin Hood'. Even after the shooting fiasco, he manages to direct (from his bed) the extinguishing of the fire for which he is responsible ('It's quite simple to put out a fire'), and takes credit for saving the family. It is, after all, easier to direct than to do.

He is the intellectual member of the family, his ambition to write, a task from which he emerges periodically to express his irritation at all distractions (a peasant's donkey, Gerry's animals, Leslie's shooting at a tin can) and to utter pungently phrased comments (on Lugaretzia, for example, or Mother's bathing-costume). He has an inexhaustible fund of words and colourful references, not all of them devoted to immortal masterpieces, for he condescends to write and deliver a funeral address at the burial of Achilles, the tortoise.

He is dogmatic and quite immovable; he is firmly convinced that the Magenpies are thieves and equally firmly convinced that Alecko is an albatross. He has a wonderful (and amusing) talent for exaggeration – he fears that Alecko's arrival will be followed by a cyclone, St Elmo's fire, a tidal wave *and* an earthquake. He pretends to be unfeeling: his comment on the story of his aunt and the bees is not that she might have been killed but that 'it completely ruined my holiday'. His impish humour is illustrated by his remarks when Kralefsky has been thrown while wrestling with Gerry.

But much of Larry's dogmatism, thoughtlessness and air of martyrdom is a pose, like his announcement that spring, for him, means death and melancholy. It is clear that his bark is worse than his bite. One must remember that he appears here only as seen through the eyes of a much younger brother.

Leslie

Short, stocky, with an air of quiet belligerence

Leslie plays a much smaller part in the story and emerges much less distinctly than Larry. He has a passion for hunting and shooting, and is very much wrapped up in his guns, with which he practises on all occasions. He turns the veranda into a shooting gallery and conducts ballistic experiments in the garden. He is particularly proud of his double-barrelled shotgun and gives the family a graphic description of how he pulled off his first 'left and right'. He becomes enthusiastic about hunting wild boar, hares and ducks, and gives an ardent account of his prowess in these pursuits, throwing himself into every part, hunter and hunted.

When Peter pays too much attention to Margo, Leslie plays the part of the outraged brother, suggests that he should shoot Peter and appears at intervals brandishing a revolver. A little later he erects a shotgun alarm for suspected thieves, which causes much panic one night. He has to put up with a good deal of banter and scorn from Larry and challenges him to come shooting snipe. In the subsequent catastrophe Leslie is much more concerned for his gun than for Larry's safety.

Margo
Trailing yards of muslin and scent

A boy of ten, his mind filled with such really important matters as animals and their habits, personalities and needs, has, naturally, little time for a sister of eighteen who is interested in diaphanous garments and books on slimming, and whose greatest worry is her acne spots. We do not therefore gain a very sharp impression of Margo. We learn of her artistic talents when she prepares for a party by drawing murals on brown paper; of her uncertainty about the finer points of idiom in English proverbs; and of her romantic feelings towards Peter, whose departure leaves her tearful and broken-hearted. She takes the part of the unhappy lover but is rather panic stricken when he threatens to return. She gives herself up to solitude, weeping and the poetry of Tennyson; nevertheless she retains her good appetite. Alone on an island, she meditates on love and, because of a sirocco, has great difficulty in getting safely home.

Spiro
Like a great, brown, ugly angel he watched over us as tenderly as though we were slightly weak-minded children.

Spiro Hakiaopulos (called Spiro Americano 'on accounts of I lives in America'), the barrel-bodied taxi driver with a great leathery face and a rich booming voice, has a great love for the English and for the Durrell family in particular. Within a week he is their guide and friendly philosopher. He finds houses for the family; installs them; protects them from officials; fetches and carries; cooks for them on picnics. He bargains for them in shops; lends them money; and helps with the preparations for their parties, obtaining the assistance of the butler of the King of Greece and even stealing goldfish from the King's pond. He adores Mother and enters fully into the emotional drama of Margo and Peter. He knows and loves everyone on the island, and 'they respected his honesty, his belligerence and above all they adored his typically Greek scorn and fearlessness when dealing with any form of governmental red tape.' He has a terrifying way of driving ('People are scarce when I drive through a village'), a great fund of invective in Greek, a strange form of English pronunciation and a complete inability to manage some English words. His gay and generous personality seems to represent the very spirit of the island of Corfu.

Theodore

No matter what the subject, Theodore could contribute something interesting to it.

Gerry first meets Dr Theodore Stephanides when he is searching for the explanation of the mysterious trapdoors he has discovered. This eccentric nature-lover makes an immediate impression upon Gerry, with his beard, his immaculate dress, his keen twinkling eyes and his inexhaustible stock of knowledge. ('He is an expert on practically everything you care to mention. And what you don't mention, he does'). His friendship with Gerry quickly ripens during their weekly meetings, from which Gerry learns a great deal of natural history. He is a successful teacher because he talks to Gerry as though he is as old and knowledgeable as himself. He makes Gerry gifts of books and a pocket microscope.

Theodore is shy, with his nervous manner of speaking and of glancing at his boots, and his habit of rasping his beard with his thumb in moments of pleasure. He takes a boyish delight in watching the weekly seaplane land in the bay and has a passion for ghost and crime stories. The book would lose much without Theodore's anecdotes about the opera and the Corfu Fire Brigade, and his awful puns ('like an ancient copy of *Punch*').

Kralefsky

I decided immediately that Kralefsky was not a human being at all, but a gnome who had disguised himself as one by donning an antiquated but very dapper suit.

Kralefsky is another eccentric, with an eccentric mother. His appearance is unusual:

He had a large, egg-shaped head with flattened sides that were tilted back against a smoothly rounded hump-back. This gave him the curious appearance of being permanently in the middle of shrugging his shoulders and peering up into the sky. A long, fine-bridged nose with widely flared nostrils curved out of his face, and his extremely large eyes were liquid and of a pale sherry colour. They had a fixed far-away look in them, as though their owner were just waking up out of a trance. His wide, thin mouth managed to combine primness with humour. (3,14,214–15)

He has three great interests in life: his birds, on which he lavishes his care and his conversation; his mother, whom he tends with constant devotion; and 'an entirely imaginary world he had evoked in his mind, a world in which rich and strange adventures were always happening'. These adventures he relates in a most dramatic and realistic manner (pp.237–40).

Kralefsky is a hard taskmaster and a boring, relentless tutor, with very old-fashioned methods. Time is something of an obsession with him, and his life is regulated by his chiming watch. He conversation (when he is not telling one of his imaginary adventures) is abrupt and bird-like and liberally sprinkled with repetitions of 'That's the ticket!' But he is always courteous and considerate. When he has been thrown wrestling with Gerry and is suffering from two cracked ribs, he remains composed and 'smiled a smile of pain-racked nonchalance ... "Please, please don't distress yourself about it ... Don't blame the boy; it was not his fault. You see, I'm a *little* out of practice."' It is Kralefsky's 'big moment'.

Achilles, the Tortoise

A most intelligent and lovable beast, possessed of a peculiar sense of humour

Gerry buys this small sprightly tortoise from the Rose-beetle Man. Achilles soon learns to answer to his name and is allowed to go where he wishes in the garden. He loves grapes and strawberries and his feeding on these is amusingly described (pp.50–1). His love of human company is rather tiresome to anyone sunbathing in the garden. He attends one of Gerry's lessons with George, but causes much trouble by getting wedged under the furniture. One day he wanders out of the garden. The family search for him, calling, 'Achilles ... strawberries'. He is found dead at the bottom of a disused well and is buried in the garden, with a funeral address written and read by Larry.

Ulysses, the Scops Owl

A bird of great strength of character, and not to be trifled with.

Gerry discovers the Scops owl while hunting for squirrel dormice and keeps him in a basket in his study. This fearless bird is introduced to Roger, who suffers from his sharp claws; later the two creatures become quite friendly towards each other. Ulysses is allowed to fly about the room and at night he goes off hunting, always coming back to the house for his supper. Sometimes he accompanies Gerry and Roger on a late evening swim, riding on Roger's back, guarding Gerry's clothes, and occasionally flying over Gerry and Roger while they are in the water.

Geronimo, the Gecko

His assault on the insect life seemed to me as cunning and well-planned as anything that famous Red Indian had achieved.

This fierce and cunning hunter does not allow any other gecko to enter Gerry's bedroom. Moving slowly across the ceiling, he stalks and then rushes upon his prey – daddy-long-legs, lacewing flies, months. He attacks any intruding gecko by seizing his tail, which he eats in triumph.

Cicely, the Mantis

Having paused, swaying from side to side on her slender legs, and surveyed me coldly, she continued on her way, mincing through the grass-stalks . . . she whirled round and stood up on end, her pale, jade-green wings outspread, her toothed arms curved upwards in a warning gesture of defiance.

Fascinated by the savage love-life of the mantids (while mating is in progress the female devours the top half of the male's body), Gerry is anxious to know about the hatching out of the eggs, so he is happy to come across a very large female mantis who seems to be 'expecting a happy event'. Cicely is an insect of great panache and courage, not afraid to take on Geronimo in a fight (3,13,204–7). For much of the time, she gives as good as she gets, but Geronimo is the victor in the end – on Gerry's bed! – when her head and thorax disappear into his mouth.

The Magenpies

Fat, glossy and garrulous... the Magenpies looked the very picture of innocence.

Gerry climbs a tree and takes two baby magpies from a nest. They receive their name because of Spiro's strange pronunciation. Larry is convinced that they will steal the family's valuables. Their first flights are described, and they begin to visit the rooms in the house. One day they carry out a raid on Larry's bedroom, with disastrous results (pp.234–5), and the Magenpies have to be confined to a cage. Here they amuse themselves, and cause confusion by imitating the voices of the family and of the maid who feeds the chickens. On the day of the 'Christmas' party, however, they escape and cause havoc with the lunch table set out on the veranda.

Old Plop, the Terrapin

He was bigger than any terrapin I had seen, and so old that his battered shell and wrinkled skin had become completely black.

Gerry is determined to capture this large creature which lives in one of the canals in the Chessboard Fields, but Old Plop, in spite of his great age, is very wily and always gets away. Eventually with the aid of Kosti, the convict, Gerry catches Old Plop and keeps him in a large stone tank (p.225).

Alecko, the Black-Backed Gull

It was like being offered an angel. A slightly sardonic-looking angel, it is true, but one with the most magnificent wings.

This magnificent bird, brought from Alabania, belongs originally to Kosti, the convict. Quite unusually, Alecko allows Gerry to stroke him without biting, and the convict says Gerry may have the bird. The boy has some difficulty in getting Alecko home (3,7,270–1), and the bird's arrival is greeted with consternation by Larry, who always refers to him as 'that albatross'. Alecko is kept in the Magenpies' cage but on the day of the 'Christmas' party he escapes and bites the guests in the leg (3,18,287–8).

Setting and structure

'This is the story,' says the author, 'of a five-year sojourn that I and my family made on the Greek island of Corfu.' This remarkable island, with its happy-go-lucky way of life, provides a most appropriate background for the story he has to tell of a lively family, whose life is full of comic incidents and astonishing alarms and excursions, and of a succession of animals who seem, as he describes them to take on human personalities, whether he observes them in their natural state or whether they are brought home to become members of the family.

Corfu is an island with ancient customs and colourful inhabitants, peasants with superstitions about kissing the feet of St Spiridion and not sleeping under cypress trees, and weird and fascinating characters like the Rose-beetle Man. The island has a varied landscape, and the setting of Durrell's story is constantly changing accordingly, as he wanders through the cicada-haunted orchards and sunken gardens of the villas which the family occupied, explores the olive groves, cyclamen woods, vineyards and cypress-shaded valleys beyond, climbs the small hills covered with myrtle and heather to the bare, rocky peaks above, picks his way among the pools and swamps of the Chessboard Fields, with their pattern of narrow waterways, or spends long hours in secluded bays with miniature cliffs and ventures out to the enchanted archipelago of small islands off the coast. Sometimes there are visits to the town of Corfu, with its cathedral, its narrow, twisted streets, its bird-market and its Jewish quarter.

The cobbled streets crammed with stalls that were piled high with gaily-coloured bales of cloth, mountains of shining sweet meats, ornaments of beaten silver, fruit and vegetables. The streets were so narrow that you had to stand back against the wall to allow the donkeys to stagger past with their loads of merchandise. It was a rich and colourful part of the town, full of noise and bustle, the screech of bargaining women, the cluck of hens, the barking of dogs and the wailing cry of the men carrying great trays of fresh hot loaves on their heads. (Part 2, Chapter 9, p.132).

Corfu is a fascinating island, and on it anything can happen.

Stressing that his anecdotes about the island and the islanders are absolutely true, Durrell says, 'Living in Corfu was rather like living in one of the more flamboyant and slapstick comic operas.' Anywhere else in the world Theodore's stories would have to be made up, 'but here in Corfu they ... er ... anticipate art, as it were'.

But, in addition to the comic episodes, *My Family and Other Animals* presents scene after scene of pure happiness, like the picnic at the lake of lilies, and dreamy afternoons when Gerry wanders over the island with Roger, exploring the marine life of the archipelago in a small boat or watching the tortoises emerging from their winter sleep.

> Gradually the magic of the island settled over us as gently and clingingly as pollen. Each day had a tranquillity, a timelessness, about it, so that you wished it would never end. But then the dark skin of night would peel off and there would be a fresh day waiting for us, glossy and colourful as a child's transfer and with the same tinge of unreality. (1,2,40)

The whole book reflects this magic – its tranquillity, its colour, its tinge of unreality and its timelessness. There is often a vivid sense of the time of day; of morning (the opening of Chapter 3); of the hot, dreamy afternoon (the opening of Ch 5); of the cool, shadowy evening (the conclusion of Ch 10). But we have only a general notion of the changing seasons from Sweet Spring to Woodcock Winter. The seasons themselves are often sensitively and vividly described (see 1,6,83), but we are aware of no detailed time-scheme as the story unfolds. It opens in August. We reach Corfu in dying summer and are told of the warm wet winter that follows. Spring comes soon afterwards (Ch 6) and merges slowly into 'the long, hot, sun-sharp days of summer' (Ch 10), which grow hotter and hotter (Ch 11). Soon there is an air of uncertainty and expectancy and winter arrives, slowly and gently (Ch 12). With another spring comes the move to the Snow-White Villa and the story concludes with the 'Christmas' party in September.

Once the family has been installed in the Strawberry-Pink Villa, the story is largely a sequence of separate incidents. It would be wrong to expect here the careful construction that goes into the writing of a novel. Occasionally there are chapters which appear to have no unity or real sense of direction: it is their separate parts that provide their interest. At the beginning of his story the author says:

> In order to compress five years of incident, observation and pleasant living into something a little less lengthy than the *Encylopaedia Britannica*, I have been forced to telescope, prune, and graft, so that there is little left of the original continuity of events. (p.9)

Some parts of the story, it is true, are governed by a real sequence of events – like the growing friendship with Theodore, or the result of Peter's attachment for Margo – and some chapters have a striking continuity as discoveries and events interact on one another. In Chapter 11 we see how the expeditions in the *Sea Cow* have aroused in Gerry a desire for a boat of his own, so that with the approach of his birthday he persuades Leslie to build him a boat; the birthday party is described, and on the following day Gerry makes his first voyage in his new boat; the chapter concludes with a description of the marine creatures which he observed on this and subsequent explorations of the islands. Chapter 11 is a tightly-knit sequence from the description of winter near the beginning to the fire in Larry's bedroom at the end. Chapter 15, too, has an interesting construction:

> It was while the dogs and I were resting after just such a hunt that I acquired two new pets, and, indirectly, started off a chain of coincidences that affected both Larry and Mr Kralefsky. (p.228)

The reader should map out for himself the whole chain of coincidences in this chapter. ('The Cyclamen Woods'). Finally, Chapter 18 provides a fitting climax to the story, for here so many animals and so many people (each one exhibited in characteristic pose and behaviour) are brought together in an atmosphere that is full of the hilarity that has permeated the rest of the book.

Style

Gerald Durrell has an enviable mastery of English prose. His style is notable for its variety.

Humour

My Family and Other Animals is a happy book. It is also a very funny book. Gerald Durrell has a keen eye for the ludicrous, which he describes with zest. Every reader will have his own favourite passages, and no one will need to have the humorous passages pointed out to him. But the student should note how many different *kinds* of humour are to be found in the book.

There are things that are funny simply because they are quite fantastic, like Kralefsky's account of his imaginary adventures.

Some of the humour springs from quirks of character, like the reasoning of the Fire Brigade Chief and Mother's constantly choosing places to be buried in. The humour of the latter is considerably enhanced by the author's comment.

> but they were generally situated in the most remote areas, and one had visions of the funeral *cortège* dropping exhausted by the wayside long before it had reached the grave. (1,6,84)

Sometimes a situation is funny because of an element of anti-climax – like the family's majestic mounting of cabs in Corfu and their eventual arrival at the Pension Suisse followed by a horde of mongrel dogs, Larry lashing out indiscriminately with a whip and Mother retaining her air of majestic graciousness to the end.

Incongruity and unexpectedness form another source of humour, as on the occasion when Leslie appears in the midst of a party, wearing only a towel, gesticulating wildly and bellowing about snakes; or, a little later, when the guests sit down to lunch and are bitten in the leg by Alecko.

There are many examples of verbal humour. Sometimes it arises from understatement and irony:

> I travelled with only those items that I thought necessary to relieve the tedium of a long journey: four books on natural history, a butterfly net, a dog, and a jam-jar full of caterpillars all in imminent danger of turning into chrysalids (1,1,18);

sometimes from overstatement or fantastic comparison:

looking like a chocolate statue that has come into contact with a blast furnace (2,12,185);

the sort of cry the minotaur would have produced if suffering from toothache (3,18,285)

and sometimes from unexpected juxtaposition:

'Don't be ridiculous, dear,' said Mother firmly; 'that's quite out of the question [i.e. selling the house]. It would be madness.'
So we sold the house (1,1,17).

Much of the humour is to be found in Larry's remarks – as in his comments on Mother's bathing-costume:

'It looks to me like a badly-skinned whale.' (2,10,151)
'It'll probably suit you very well if you can grow another three or four legs to go with it.' (2,11,152)
'like a sort of marine Albert Memorial.' (2,11,153)

An incident often owes its humour to the manner in which it is told. This applies particularly when animals are involved. The account of the Magenpies' raid on Larry's room, for example, is funny because of the way in which the birds' supposed motives and their behaviour are described; from the description of the havoc in the room; and, of course, from Larry's comments. Often the humour of an incident depends for a great deal of its effect on the cool and detached manner in which it is told – as Larry's story of the aunt who was nearly stung to death by bees:

'To be drenched with cold water and then hit on the head with a large galvanized-iron bucket is irritating enough, but to have to fight off a mass of bees at the same time makes the whole thing extremely trying.' (3,18,290)

Sometimes a dramatic or near-tragic situation is made to appear funny by the comments and remarks of the participants who are acting at cross-purposes. Undoubtedly the funniest example is the Thurberesque episode when Leslie's shotgun alarm goes off in the night and causes a panic (Ch 12).

Finally there are examples of purely physical humour of the slapstick kind. One example is the episode of the female scorpion who, released from the matchbox by Larry, scatters her babies all over the dining-table to cause pandemonium, in which

Mother is drenched in cold water and Lugaretzia bitten in the ankle by Roger. Another example is Larry's ill-fated attempt to shoot snipe, ending in his fall in the mud. The irate conversation of the two brothers adds much to the humour of this incident and the subsequent episodes, when Larry is drunk and causes a fire, and Mother appears, 'struggling, for some reason best known to herself, to get her corsets on over her nightie' (2,12,187).

Narrative

He writes of the ludicrous with great gusto, as when he describes the family's undignified arrival at the Pension Suisse (Ch 1). At one moment he gives us an easy, unstrained account of activities on 'hot, dreamy afternoons' (Ch 5); at another he writes in slow and deliberate sentences, as in his exact relation of a tortoise's emergence from its hibernation (Ch 8). The student should contrast the rather elaborate sentence construction and careful choice of words in the account of Mother's bathing (Ch 10) with the speed and bustle of the passage in Chapter 11 which tells of the guests dancing the *Kalamatiano*. The student will be able to find other examples of Durrell's narrative skill.

Conversation

Much of the book consists of lively conversation (this is not, of course, confined to the author's two sections entitled 'Conversation'). Often it is conversation in which the whole family participates, each member contributing rapid and characteristic remarks (as in the episode of the scorpion which scatters baby scorpions over the dining-table); and sometimes it is the irate interchange of two brothers (as in the conversation of Larry and Leslie in Ch 11, pp.162–3). The student should also note the way in which Durrell reproduces the style of individuals – Spiro's breathless, machine-gun-like rattle; Kralefsky's bird-like trilling, Theodore's nervous hesitancy and careful choice of words. Sometimes Durrell obtains a special effect, suggesting a rather exaggerated and self-conscious harmony between the speakers, by putting a whole conversation into reported speech, as in the discussion about the boat between Leslie and Gerry (2,11,160–1).

Description

Here again Durrell shows great skill and variety.

People His descriptions of people show them all as colourful, extraordinary or eccentric – from the Rose-beetle Man to the Belgian consul. Many of the important characters are dealt with in the previous section of these notes, p.35; but the student should notice the skill with which some of the very minor characters are presented in thumb-nail sketches. Examples are: George:

> George was a very tall and extremely thin man who moved with the odd disjointed grace of a puppet. His lean, skull-like face was partially concealed by a finely pointed brown beard and a pair of large tortoise-shell spectacles. He had a deep, melancholy voice, a dry and sarcastic sense of humour. Having made a joke, he would smile in his beard with a sort of vulpine pleasure which was quite unaffected by anyone else's reaction. (1,4,57)

Lugaretzia:

> she was a thin, lugubrious individual, whose hair was forever coming adrift from the ramparts of pins and combs with which she kept it attached to her skull. (2,7,100)

and Dr Androuchelli:

> a little dumpy man with patent-leather hair, a faint wisp of moustache, and boot-button eyes behind great horn-rimmed spectacles. (2,7,105)

Insects, birds and flowers The same technique enables Durrell to give us the most vivid and striking descriptions of insects:

> Among the myrtles the mantids moved, lightly, carefully, swaying slightly, the quintessence of evil. They were lank and green, with chinless faces and monstrous globular eyes, frosty gold, with an expression of intense, predatory madness in them. The crooked arms with the fringes of sharp teeth, would be raised in mock supplication to the insect world, so humble, so fervent, trembling slightly when a butterfly flew too close. (2,10,141)

of birds:

> Its squat beak, with a yellow fold at each corner, the bald head, and the half-open and bleary eyes gave it a drunken and rather imbecile look. The skin hung in folds and wrinkles all over its body, apparently pinned loosely and haphazardly to its flesh by black feather-stubs. Between the lanky legs drooped a huge flaccid stomach, the skin of it so fine that you could dimly see the internal organs beneath. The baby

squatted in my palm, its belly spreading out like a water-filled balloon, and wheezed hopefully. (3,15,230)

and of flowers:

Roses dropped petals that seemed as big and smooth as saucers, flame-red, moon-white, glossy, and unwrinkled; marigolds like broods of shaggy suns stood watching their parent's progress through the sky. In the low growth the pansies pushed their velvety, innocent faces through the leaves, and the violets drooped sorrowfully under their heart-shaped leaves. The bougainvillaea that sprawled luxuriously over the tiny front balcony was hung, as though for a carnival, with its lantern-shaped magenta flowers. In the darkness of the fuchsia-hedge a thousand ballerina-like blooms quivered expectantly. (1,2,30)

Animals But it is in his descriptions of animals that Durrell is most successful. There are so many of these that the reader will have his own favourites. He should study them in detail, noting the adjectives, the striking comparisons, the selection of verbs and the touches of humour. He should also memorize some of the details of his chosen passages; and this applies also to the more important animals considered in the previous section of these notes. Durrell is equally successful with such different types of description as those of the sea creatures round the islands of the enchanted archipelago and the almost human conversational qualities of Roger in Chapter 3. Although his descriptions of animals are often of whole species and deal with their typical and characteristic features, Durrell often gives us a careful and detailed study of a single creature, like the pages devoted to Geronimo, the gecko, in Chapter 13, or the description of the two toads in the same chapter.

Each one had a girth greater than the average saucer. They were greyish-green, heavily carunculated, and with curious white patches here and there on their bodies where the skin was shiny and lacking in pigment. They squatted there like two obese, leprous Buddhas, peering at me and gulping in the guilty way that toads have. Holding one in each hand, it was like handling two flaccid, leathery balloons, and the toads blinked their fine golden filigreed eyes at me, and settled themselves more comfortably on my fingers, gazing at me trustfully, their wide, thick-lipped mouths seeming to spread in embarrassed and uncertain grins. (p.209)

Here the student will notice the exact vocabulary ('carunculated', 'flaccid', 'filigreed'), the careful description of colour, the comparison ('like two obese, leprous Buddhas'), and Durrell's trick of attributing human motives and feelings to the animals he describes.

This is a characteristic feature of his writing and occurs so often that an unsympathetic critic might object to it as overdone and perhaps even protest, as Larry does on one occasion, about 'anthropomorphic slush'. A dog has 'a ridiculous grin'; a tortoise has 'an expression of bemused good humour on his face': a grasshopper has 'a long melancholy face'; dead turtle-doves have 'demurely closed eyes'; tortoises wander about 'with an air of preoccupied determination' – and so on. Occasionally, it seems, his beloved creatures let him down by refusing to behave like human beings – as the scorpions, of whom he says, 'Several times I found them eating each other, a habit I found most distressing in a creature otherwise so impeccable.' To the objection that this is not a scientific approach to animals the answer can be made that *My Family and Other Animals* is not a scientific textbook; in the author's own phrase, it is 'a *mildly nostalgic* account of the natural history of the island'. It is the record of a world and a half-decade seen through the eyes of a child, and a child sharing to the full all the natural delights of a Mediterranean island. Animals were his love and many of them were his pets. It is therefore only to be expected that he should see them through very sympathetic human eyes.

Colour, scent and sound Durrell possesses to a high degree the ability to recreate vividly the sensual impressions of the world around him. His descriptions are always bright with colour. Two passages, in particular, repay study in this respect. The first (pp.18–19) is the description of the island as he first saw it from the boat, which begins 'The sea lifted smooth blue muscles of wave'. The second passage is the opening of Chapter 5 (p.66). The student should read both passages carefully, then decide which one he prefers.

He should also notice references to the sense of smell in Durrell's writing; for example

a thousand white flowers in the sunshine like a multitude of ivory horns lifting their lips to the sky and producing, instead of music, a rich, heavy scent that was the distilled essence of summer, a warm sweetness that made you breathe deeply time and again in an effort to retain it within you. (3,16,256)

Style

And the vivid descriptions of sounds:

the quick *wheep* of wings (1,6,9)
chaffinches... *pinking* like a hundred tiny coins (1,6,91)
a bat... *chittering* with dark malevolence (3,13,199)

Often, impressions of colour, scent and sound are combined in a single passage.

In the morning, when I woke, the bedroom shutters were luminous and barred with gold from the rising sun. The morning air was full of the scent of charcoal from the kitchen fire, full of eager cock-crows, the distant yap of dogs, and the unsteady, melancholy tune of the goat-bells as the flocks were driven out to pasture. (1,3,41)

Vivid detail The book contains many instances of the way in which the author gives the reader distinct and realistic impressions by means of carefully observed and selected details.

The dark waves lifted our wake and carried it gently towards them, and then, at their very mouths, *it crumpled and hissed thirstily among the rocks.* (1,1,18–19)
Theodore would... give me his characteristic handshake – *a sharp downward tug, like a man testing a knot in a rope.* (1,6,80)
Tea would arrive, the cakes *squatting on cushions of cream*, toast *in a melting shawl of butter*, cups agleam, and a faint wisp of steam rising from the teapot spout. (1,6,82)
The *beetle-shiny* barrels [of a shotgun] (1,6,91)
Dogs' faces [seen from high up a tree] *the size of pimpernel flowers.* (3,15,129)

Lively comparisons Like all Durrell's books, *My Family and Other Animals* is a treasury of the sort of comparison that comes from a vivid imagination allied to the eye of a naturalist or a painter. Sometimes they are brief but evocative.

Green shutters folded back from the windows... of multi-coloured houses... like the wings of a thousand moths. (1,1,20)
Lady-birds moved like newly painted toys. (1,2,36)
Linnets in their neat chocolate-and-white tweed suiting. (3,14,217)

Often they are exact and detailed.

(Of cyclamens) a fountain of beautiful flowers that looked as though they had been made from magenta-stained snowflakes. (3,15,228)
In a few days small white clouds started their winter parade, trooping across the sky, soft and chubby, long, languourous, and unkempt, or small

and crisp as feathers, and driving them before it, like an ill-assorted flock of sheep, would come the wind. (2,12,177)

Atmosphere and imagination Durrell is just as successful in painting on a broader canvas and in creating a general *atmosphere*. Some of the best examples of this skill are his descriptions of the different seasons of the year. The opening paragraph of 'The Migration' (p.15) is such an example. There is a striking paragraph in Chapter 6 (p.83) which describes the coming of spring to the island, the change that comes over the cypresses and the olives, the appearance of flowers, and the effect of all this on both human and animal life. A shorter paragraph on spring occurs at the beginning of the second 'Conversation' (p.190)

Another little masterpiece is the description of summer in the first four paragraphs of Chapter 10 (pp.141–2). But undoubtedly the finest and most sustained of all these evocations of the seasons is that of the gentle coming of winter in Chapter 12 (pp. 176–8), beginning with uncertainty and the small signs suggesting that nature is preparing for something new.

This passage also reveals the vividness of Durrell's *imagination* – a quality illustrated again and again in the book. Consider, for example, how young Gerry is haunted by the idea of the trap-door spider crouching in its silken tunnel, listening to the movements of the insects outside.

I could imagine that a snail would trail over the door with a noise like sticking-plaster being slowly torn off. A centipede would sound like a troop of cavalry. A fly would patter in brisk spurts, followed by a pause while it washed its hands – a dull rasping sound like a knife-grinder at work. The larger beetles, I decided, would sound like steam-rollers, while the smaller ones, the ladybirds and others, would probably purr over the moss like clockwork motor-cars. (1,5,78)

Again the boy's imagination is stirred by the eccentric behaviour of the Belgian consul who teaches him French.

For the rest of the morning I toyed with the exciting idea that the consul had committed a murder before my very eyes, or, at least, that he was carrying out a blood feud with some neighbouring householder. But when, after the fourth morning, the consul was still firing periodically out of his window, I decided that my explanation could not be the right one, unless it was an exceptionally large family he was feuding with, and a family, moreover, who were apparently incapable of firing back. (2,9,134)

The author's childhood view of things is evident when he hears Kralefsky's watch for the first time.

> To my astonishment the noise appeared to emanate from somewhere inside Kralefsky's stomach... He inserted finger and thumb into his waistcoat and drew out his watch. He depressed a tiny lever and the ringing sound ceased. I was a little disappointed that the noise should have such a commonplace source; to have a tutor whose inside chimed at intervals would, I felt, have added greatly to the charm of the lessons. (3,14,218)

My Family and Other Animals is written with the artistry of a mature author, but much of its charm lies in its recreation of a world seen through the eyes of childhood.

General questions and questions on related topics for coursework/examination for other books you may be studying

1 Compare the methods employed and the success achieved by Gerry's four tutors. (Note: Theodore was not one of his tutors.)

Suggested notes for essay answer:

George. In the absence of school books he made use of books from his own library – *Geography* done from maps in Pears Cyclopaedia (copies filled in, not with national products, but with native animals); *English* from very adult classics; *French* from an illustrated dictionary; *Mathematics* 'from memory'. While Gerry worked George practised fencing and peasant dances. George instituted lessons out of doors; got Gerry interested 'by seasoning a series of unpalatable facts with a sprig of zoology' – not always authentic. In natural history lessons he made Gerry realize importance of careful observation and the keeping of records; thus Gerry learned in a systematic way and remembered much more.

The Belgian Consul. In French lessons Gerry had to read from a French dictionary and the consul fired an air-rifle through the window at cats. Gerry learned no French; the morning lessons so boring that afternoon natural history sorties were subsequently carried out with extra enthusiasm.

Peter. Gerry found his strict approach very trying at first. Later some lessons while swimming along the coast. Each day Gerry practised English by writing a chapter of a sensational story while Peter spent much time with Margo. Mother dispensed with his services, fearing he was becoming too fond of Margo.

Kralesfsky. As odd-looking little man with an attic full of caged birds. A stickler for work; lessons were very boring – *history* meant learning dates; *geographical facts* to be learnt by heart; *French* from a three-volume set of bird books. Kralefsky took up much of their time recounting fantastic adventures with his many female acquaintances. In a wrestling lesson he crashed to the floor, cracked two ribs.

Kralefsky and the Belgian consul taught Gerry little; Peter encouraged him to write imaginative English; George was probably the most successful of his tutors, especially in the matter of natural history.

2 '*My Family and Other Animals* is a *happy* book.' Illustrate this statement by close reference to some of the episodes in the story.

3 Write an account of *either* 'The Enchanted Archipelago' *or* 'The Woodcock Winter', bringing out carefully the sequences of events in the chapter you have chosen.

4 Write an account of the events described in 'An Entertainment with Animals', showing how this chapter knits together characters and animals which have appeared earlier in the story.

5 Discuss the humour of the book, with close reference to carefully chosen examples.

6 Show how *either* Larry *or* Kralefsky contributes to the humour of the book.

7 'Most of Durrell's characters are eccentrics.' Discuss this statement, with close reference to some of the portraits of people which the book contains.

8 Give an account of *either* the birds *or* the insects which appear in the book, showing how you have been impressed by Durrell's vivid description of them.

9 By means of carefully chosen examples show how Durrell attributes human motives and feelings to the creatures he describes.

10 Discuss and illustrate Durrell's use of vivid detail in his descriptions.

11 'A gentle and understanding Noah.' Show how this description fits the character of Gerry's mother.

12 'Larry was always full of ideas about things of which he had no experience.' Illustrate this statement by reference to some of the episodes in which Larry appears.

13 Give an account of the part played in the story by Spiro.

14 Write a character-sketch of Theodore.

General questions

15 Compare and contrast Theodore and Kralefsky.

16 Who was Kosti Panopoulos? Give an account of the part he plays in the story.

17 Describe Gerry's visit to Mrs Kralefsky. What subjects did they talk about?

18 Say what you have learnt from *My Family and Other Animals* about the island of Corfu.

19 Discuss the author's account of the Magenpies and show how he describes them as though they had human thoughts and motives.

20 Describe the appearance and habits of Ulysses, the Scops owl.

21 Describe Alecko, the black-backed gull. How did Gerry come by him? Give an account of the bird's later activities.

22 Say what you know about Achilles, the tortoise, and Geronimo, the gecko.

23 What does the author tell us about (*a*) tortoises, and (*b*) toads?

24 What have you learned about spiders from this book?

25 Say what you know about each of the following:
St Spiridion, Christian Science, Hercules, mantids, Perseus, Neanderthal man.

26 Which of Gerry's animals do you think you would have felt most attached to, and why?

27 Gerry's Mother is continually being faced with difficult or embarrassing situations. Choose three of these occasions. Explain the difficulties and say how she coped with them.

28 'Living in Corfu was rather like living in one of the more flamboyant and slapstick comic operas.' Illustrate the truth of Durrell's statement by referring to episodes in *My Family and Other Animals*.

29 Write an appreciation of any book you have read in which an animal or animals occupy most of your interest.

My Family and Other Animals

30 Describe a relationship between a man/woman and his/her pets in a book you have studied.

31 Write about any book you have read where animals are given human characteristics.

32 Give a detailed account of any *two poems* you have read which deal with an aspect of wild life.

33 Write about a character in any book you know who has spent much time trying to understand or care for wild life.

34 Write a story or a poem about a creature, taking care to describe it accurately and imaginatively.

35 Write about any book you have read where either cruelty to wild creatures *or* hunting or capturing plays a major part in the action.

Further reading and reference

Gerald Durrell, *Birds, Beasts and Relatives* (Fontana/Collins, 1971)
A sequel to *My Family and Other Animals*.

Gerald Durrell, *The Garden of the Gods* (Fontana/Collins, 1980) A further sequel.

Gerald Durrell, *The Amateur Naturalist* (Hamish Hamilton, 1982) This book has several paragraphs which deal with young Gerry's animal collecting on Corfu.

Lawrence Durrell, *Prospero's Cell – a guide to the landscape and manners of the island of Corcyra* [*i.e.* Corfu] (Faber, 1945)

There is a large-format Penguin edition of *My Family and Other Animals*, illustrated with Gerald Durrell's own drawings.

Grafton Books have published a more expensive illustrated edition of *My Family and Other Animals*, with line drawings by the author, black and white original photographs and specially commissioned colour pictures by Peter Barratt.

There is a Spoken Word cassette of Gerald Durrell reading *My Family and Other Animals*, LfP 7318.

Nigel Davenport reads *The Picnic and Suchlike Pandemonium* on two cassettes in Chivers Audio Books, CAB 2119–20.